82 Topics in Current Chemistry
Fortschritte der Chemischen Forschung

Large Amplitude Motion in Molecules II

Springer-Verlag
Berlin Heidelberg GmbH 1979

This series presents critical reviews of the present position and future trends in modern chemical research. It is addressed to all research and industrial chemists who wish to keep abreast of advances in their subject.

As a rule, contributions are specially commissioned. The editors and publishers will, however, always be pleased to receive suggestions and supplementary information. Papers are accepted for "Topics in Current Chemistry" in English.

ISBN 978-3-662-15422-9 ISBN 978-3-540-35249-5 (eBook)
DOI 10.1007/978-3-540-35249-5

Library of Congress Cataloging in Publication Data. Main entry under title: Large amplitude motion in molecules I-II. (Topics in current chemistry; 81-82. Bibliography: p. Includes index. 1. Molecular structure – Addresses, essays, lectures. I. Series. QD1.F58 · vol. 81 [QD461] · 340'.8s [541'.22] · 79-4221

Originally published by Springer-Verlag Berlin Heidelberg New York in 1979
Softcover reprint of the hardcover 1st edition 1979

2152/3140–543210

Contents

Low-Frequency Vibrations in Small Ring Molecules

Lionel A. Carreira

Department of Chemistry, University of Georgia, Athens, Georgia 30602, U.S.A.

Richard C. Lord

Spectroscopy Laboratory, Massachusetts Institute of Technology, Cambridge, Massachusetts 02139, U.S.A.

Thomas B. Malloy, Jr.

Department of Physics, Mississippi State University, Mississippi State, Mississippi 39762, U.S.A.

Table of Contents

I Introduction

It is well known that the mean-square amplitude of a molecular vibration in a given quantum state is inversely proportional to the product of the reduced mass and the frequency of the vibration. If it has both a small reduced mass and a frequency that is low because of small force constants, the resultant amplitude of vibration will be large. Among the modes of vibration that have small force constants are the inversion of pyramidal molecules, torsion about single bonds, bending of quasi-linear molecules, and the vibrations of ring molecules parallel to the axis of the ring. All but the last are discussed in other chapters of this volume; here we will be concerned with small ring molecules composed of light atoms, whose ring puckering vibrations have large amplitudes and hence cannot be treated by the addition of small terms of higher order to a quadratic potential system.

Transitions between the vibrational states of low-frequency vibrations are directly observable in the far infrared absorption spectrum or as low-frequency Raman shifts. The Boltzmann factors of these states are relatively large at room temperatures, even for fairly high values of the vibrational quantum number v. In addition the transition moments are substantially larger than those of a harmonic oscillator for the same v's, and thus it is often possible to observe a long progression of transitions either in infrared absorption or the Raman effect or both. When such observations extend to levels above a barrier to inversion, the potential-energy curve or surface can be mapped with accuracy to points higher than the barrier and the barrier determined with an accuracy corresponding to that of the spectroscopic measurements.

Since the potential-energy function for low-frequency vibrations involves weak force constants, the function is sensitive to intermolecular forces, which can reach comparable magnitudes to the intramolecular ones at short intermolecular distances. Thus in the liquid states the intramolecular levels are so seriously broadened as to make them difficult to observe, while in crystals the inversion barriers are drastically altered. Thus the most meaningful spectra are necessarily observed in the gas phase, and this delayed the development of the subject until suitable far infrared, laser Raman and microwave techniques were developed, as summarized below.

The experimental observation of ring-puckering motions was preceded by some insightful theoretical suggestions. Bell[1)] recognized that the ring-puckering vibration in cyclobutane should have a large quartic term in its potential function, and Rathjens et al.[2)] proposed a double-minimum potential for this molecule which predicted a highly anharmonic set of levels for the ring puckering. Pitzer and co-workers[3)] also were led by studies of the heat capacity and entropy of cyclopentane to postulate an unusual relationship between the two components of the approximately degenerate ring-puckering vibrations in that molecule. This relationship they termed "pseudo-rotation" because the successive displacements in the vibration give the appearance of a rotation of the distorted molecule.

The first direct spectroscopic observation of the highly anharmonic nature of ring puckering in a small ring molecule was made by Danti[4)], who found part of the progression for oxetane at 90–140 cm^{-1} in far infrared absorption. This work was confirmed by the spectrum of oxetane-d_6[5], and the nature of the isotopic shifts as well as the absolute values of the frequencies made it clear that the potential function is nearly a pure quartic. Pseudorotation was not observed first in cyclopentane but in the analog oxolane (tetrahydrofuran)[6]. In the latter molecule there is a small barrier to pseudorotation, as is discussed in Section IV. B., below. Quantitative measurement of the pseudorotational energy levels in cyclopentane, which confirmed the postulates of Pitzer et al.[3], was first made by means of combination bands in the mid-infrared by Durig and Wertz[7].

Since this early work the theoretical and experimental aspects of low-frequency ring vibrations have grown rapidly. In the sections below we sketch briefly the experimental methods developed to investigate ring molecules (Section II), review the theoretical basis for the interpretation of the spectroscopic data (Section III), and give an illustrative survey of the applications of the theory to the special cases of individual molecules (Section IV).

II Experimental Methods

A. Far and Mid Infrared Spectroscopy

Since the early far infrared work on ring molecules[4, 5] there has been a considerable improvement in instrumentation, first in conventional grating spectrometers[8] and

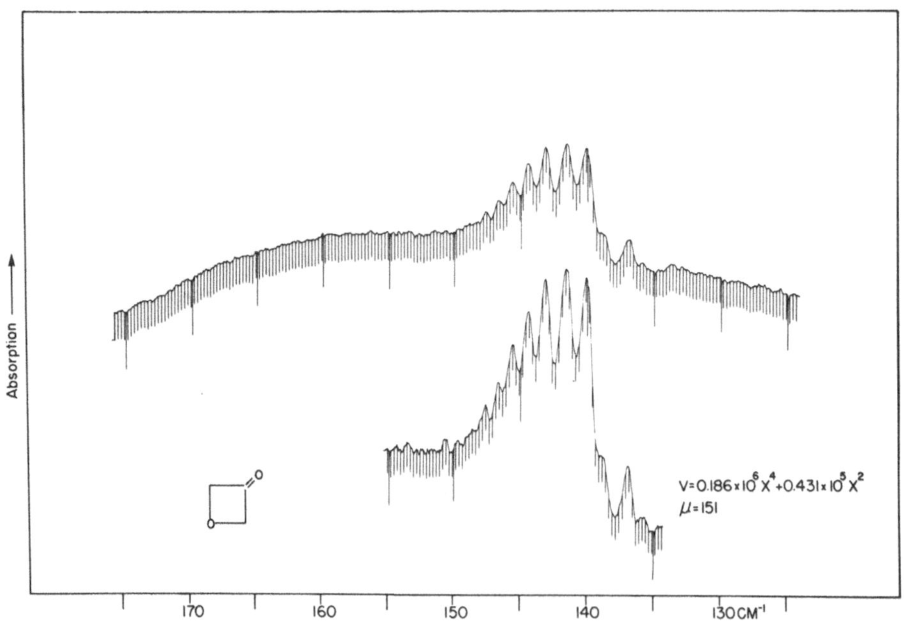

Fig. 2.1. Far infrared spectrum of oxetanone-3. The Q-branch transitions are shown on the background of overlapped P and R transitions. P = 22 torr, pathlength = 30 cm. Absorption is plotted upward in this spectrum.
[Reproduced from Carreira, L. A., Lord, R. C.: J. Chem. Phys. *51*, 3225 (1969).]

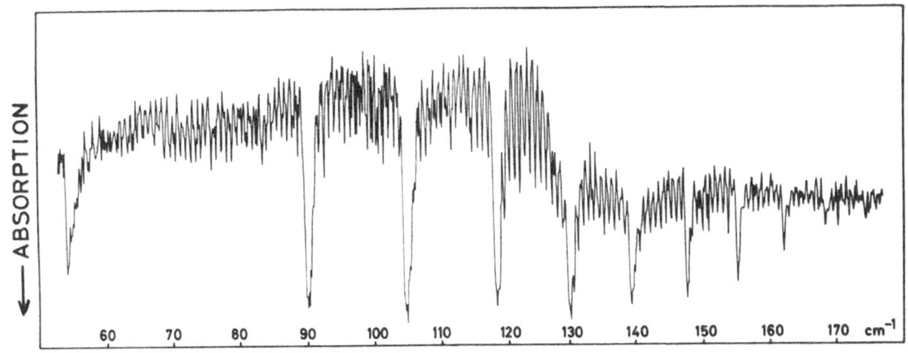

Fig. 2.2. Far infrared spectrum of oxetanone determined interferometrically. P = 48 torr, pathlength = 1 m. Absorption is plotted downward in this spectrum.
[Reproduced from Jokisaari, J., Kauppinen, J.: J. Chem. Phys. *59*, 2260 (1973)]

later in Fourier-transform interferometers[9]. Figure 2.1 shows the spectrum of oxetanone-3 recorded with a grating spectrometer[10] and Fig. 2.2 depicts the spectrum of oxetane as computed from the scans of an interferometer[9]. Both these spectra represent great improvements over those obtained in earlier studies[5] and the quantitative interpretation of the spectra is correspondingly improved. For the details of far-infrared instrumentation and techniques, the reader is referred to the monograph of Moeller and Rothschild[11].

In the mid infrared region, ring-puckering vibrations may be seen as combination and difference bands with another normal mode of vibration, as was first shown by Ueda and Shimanouchi[12]. The combination and difference bands for cyclobutane between the ring-puckering mode and the B_2 deformation frequency near 1450 cm^{-1} are illustrated in Fig. 2.3, taken from Miller and Capwell[13]. All of the combination and difference transitions (except the first difference line) involve ring-puckering energy levels in both the excited state and the ground state of the mid infrared normal mode. Since the ring-puckering energy levels in the excited state of the mid infrared normal mode may be different from those in the ground state, appropriate differences between the combination bands and difference bands must be taken to

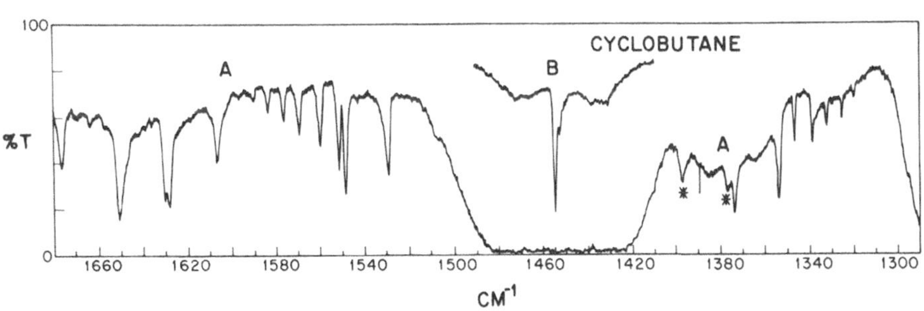

Fig. 2.3. Combination and difference band progressions involving the ring puckering vibration and a CH_2 scissoring mode in the mid-infrared spectrum of cyclobutane.
[Reproduced from Miller, F. A., Capwell, R. J.: Spectrochim. Acta *27A*, 947 (1971).]

obtain ring-puckering energy level differences between levels of the same parity in the ground state of the mid infrared mode. The symmetry of the mid infrared mode should be such that the combination and difference bands yield type-c band contours. For most C_{2v} molecules the B_2 ring-puckering mode will yield type-c sum and difference bands when the mid infrared reference band has symmetry A_1. The reference band is usually seen as a polarized fundamental in the Raman spectrum. Combination and difference bands with fundamentals of other symmetries are usually too weak or diffuse to be seen. Interpretation of the sum and difference bands can often be complicated by many other weak bands in the same region. For this reason the combination-difference band technique is usually used when no other technique is available to obtain the values of the ring puckering energy levels.

B. Raman Spectroscopy

With recent advances in laser technology, Raman spectroscopy has become a powerful tool for the direct observation of ring-puckering vibrational frequencies. Since the Raman signal in gases is extremely weak with respect to the exciting line, the use of high powered lasers, monochromators with low stray light, and efficient detection systems is necessary. The ring-puckering transitions with $\Delta v = 1$ are usually not totally symmetric and do not give rise to sharp Q branches in the Raman spectrum; on the contrary, the Raman lines of these non-totally symmetric vibrations tend to have broad contours with no discernible fine structure. However, the Raman spectra of small ring compounds show unexpected selection rules due to the very large amplitude of the

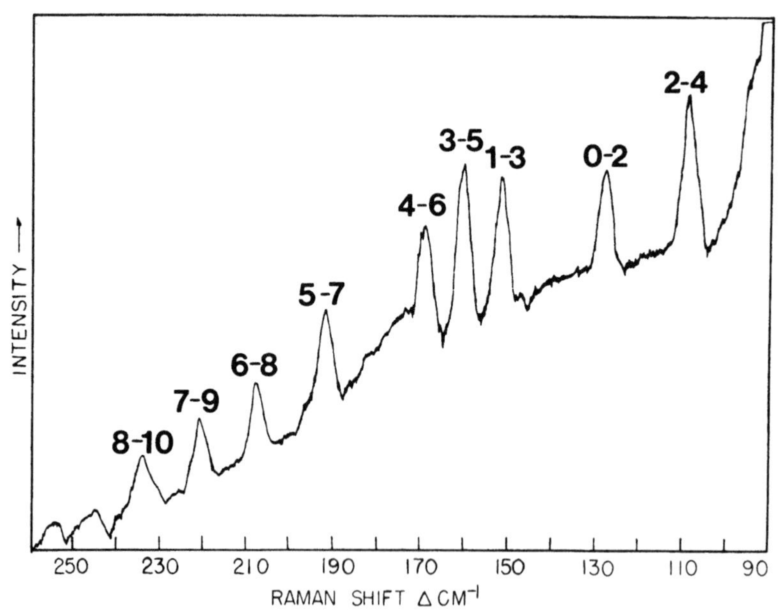

Fig. 2.4. Raman spectrum of cyclopentene.
[Reproduced from Chao, T. H., Laane, J.: Chem. Phys. Lett. *14*, 595 (1972).]

ring-puckering vibrations. Both mechanical anharmonicity and electrical anharmonicity allow overtones of the ring-puckering vibration ($\Delta v = 2$) to be observed even when the transitions with $\Delta v = 1$ are too weak or diffuse to be observed. The overtone transitions are always totally symmetric (or have a totally symmetric component) and are Raman allowed. Since the isotropic polarizability terms are nonzero and usually larger than the anisotropic, a sharp Q branch structure will be observed for the overtones and this allows the individual hot bands to be assigned. The overtone vibrations are farther removed from the exciting line so that interference from the Rayleigh line and the pure rotational envelope is minimized.

Figure 2.4 shows the Raman spectrum of cyclopentene vapor[14)] in which all the Q branches correspond to transitions with $\Delta v = 2$. Many transitions with $\Delta v = 1$ fall in this same spectral range but are too weak or diffuse to be identified.

Since C-H stretching vibrations are very strong in the Raman spectra of compounds containing C-H groups, the combination-difference band technique for determining the ring-puckering energy levels may be of use when the laser power is not sufficient for direct observation of the overtone transitions. Here the direct product of the symmetry of the C-H fundamental and the ring-puckering fundamental must belong to the totally symmetric representation. For the most common C_{2v} case the symmetry of the ring-puckering vibration is B_2 and therefore the ring-puckering sum and difference bands will have sharp Q branches when the reference C-H stretching mode has symmetry B_2. The reference band will show a sharp Q branch in the infrared spectrum but will usually be very weak or missing in the Raman spectrum. The use of sum and difference combinations allows one to obtain ground state separations between the ring-puckering levels of the same parity[12, 13)].

C. Microwave Spectroscopy

The analysis of the microwave rotational spectra of small ring compounds can provide valuable information about the nature of large-amplitude ring-puckering potential functions. The type of information obtained may vary, depending on the potential function.

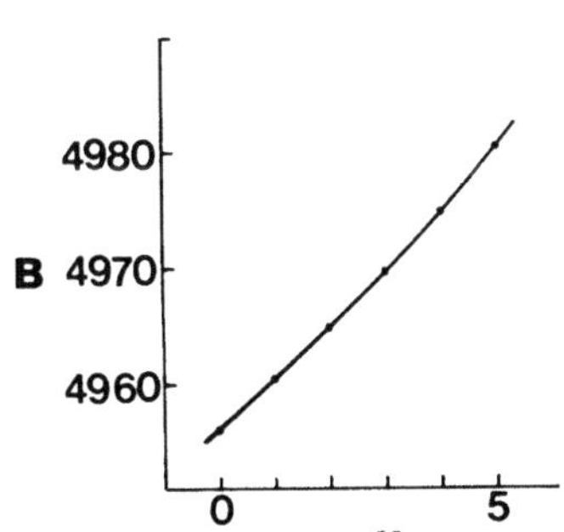

Fig. 2.5. Variation of the B rotational constant (in MHz) with ring-puckering vibrational state for oxetanone-3. Similar variations are found for the A and C rotational constants. [Reproduced from Gibson, J. S., Harris, D. O.: J. Chem. Phys. 57, 2318 (1972).]

Due to the large amplitude of vibration, a significant contribution to the effective rotational constants may be made by the ring-puckering vibration. For the simplest case, that of a molecule with a single-minimum potential function, the rotational spectra for low J transitions may be fitted with a rigid-rotor Hamiltonian. Since the vibrational frequency is low, rotational transitions in a number of excited states of the ring-puckering vibration may be observed. As opposed to the linear dependence on vibrational state expected for a small amplitude vibration, the dependence for a large-amplitude mode may exhibit curvature. This is shown in Fig. 2.5 for oxetanone-3[15]. The curvature yields information on the aharmonicity of a single-minimum potential function.

The variation of rotational constants with ring-puckering vibrational state is very sensitive to the presence of a barrier at the planar conformation. This is shown for cyclobutanone[16] and methylenecyclobutane[17] in Fig. 2.6. The presence of a very small barrier, ca. 7.6 cm^{-1} in the case of cyclobutanone, causes deviation from a smooth variation for the lower levels. In the case of methylenecyclobutane, a very pronounced zig-zag of the rotational constants is observed due to the presence of a 140 cm^{-1} barrier. The dependence of the rotational constants on vibrational state may be used quantitatively to determine the shape of the potential function as discussed in subsequent sections.

In addition, for molecules with double-minimum potential functions where vibrational levels coalesce to form inversion doublets, the microwave data may yield very accurate values for these small vibrational spacings. The microwave rotational spectrum in these states may deviate significantly from that expected for a rigid-rotor model. Since the vibrational energy spacing for this pair of levels is no longer much greater than rotational energy spacings, it is not always possible to separate the vibrational and rotational Hamiltonians. An energy level diagram for the inversion doublet in trimethylene sulfide[18] is shown in Fig. 2.7. The vibrational spacing,

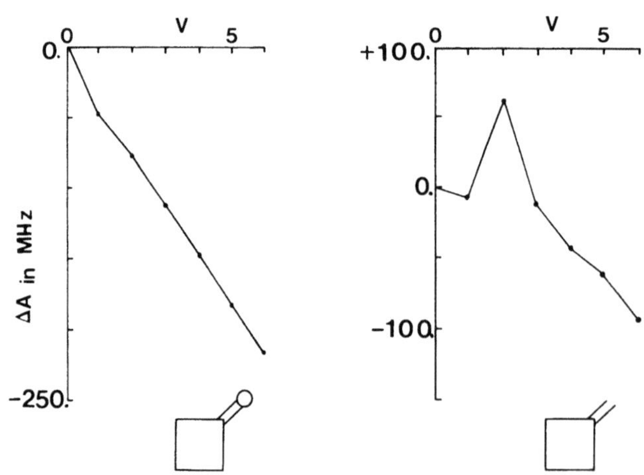

Fig. 2.6. Variation of the A rotational constants (in MHz) with ring-puckering vibrational state for cyclobutanone and methylene-cyclobutane.
[Reproduced from (A) Scharpen, L. H., Laurie, V. W.: J. Chem. Phys. *49*, 221 (1968); (B) J. Chem. Phys. *49*, 3041 (1968).]

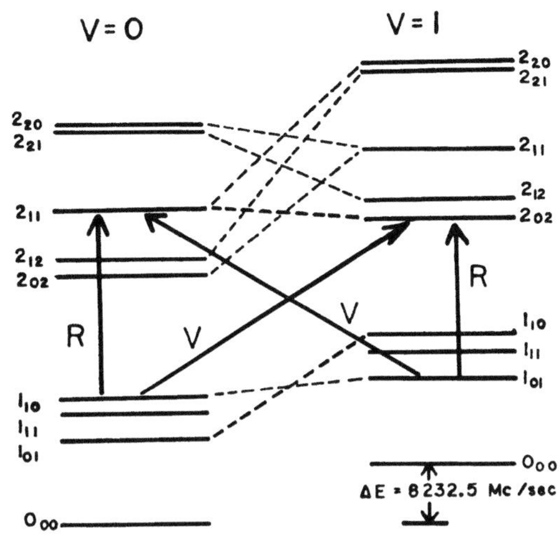

Fig. 2.7. Interactions between rotational levels in the v = 0 and v = 1 states of trimethylene sulfide.
[Reproduced from Harris, D. O., Harrington, H. W., Luntz, A. C., Gwinn, W. D.: J. Chem. Phys. *44*, 3467 (1966).]

8232.5 MHz (ca. 0.27 cm^{-1}) is quite small, in the microwave region. The interactions between rotational levels which are allowed by symmetry are shown by dashed lines.

A recent review of ring-puckering vibrations with emphasis on the theory and applications of microwave spectroscopy has appeared. The reader is referred to this work for further details[19].

III Theoretical Basis for Interpretation of the Spectra

A. Introduction

In this section, we outline a procedure for obtaining a Hamiltonian for the treatment of low-frequency vibrations in molecules. We do this, in particular, to point out the justification for some of the Hamiltonians used in the past and to make clear the nature of the approximations involved in arriving at a specific Hamiltonian. Since there is danger of overinterpreting the results obtained from approximate Hamiltonians, we indicate some of the pitfalls in doing so.

The emphasis in this section is on the form of the kinetic energy operator. The choice of potential energy functions is considered in Section IV dealing with specific molecules. Emphasis is also placed on treatment of the vibrational data since a review emphasizing treatment of microwave data for ring puckering has appeared recently[19].

We start by writing the classical kinetic energy expression for a non-linear molecule in a center-of-mass coordinate system. No distinction is made between small and large vibrational coordinates at this stage. We then rewrite the expression in a form suitable for obtaining the quantum mechanical kinetic energy operator. At that point, approximations are made to simplify the Hamiltonian and then we proceed to the treatment of the large-amplitude ring vibrations in molecules.

B. The General Hamiltonian

The kinetic energy for a molecule in a center-of-mass coordinate system may be written as

$$2T = (\boldsymbol{\omega}^t, \dot{q}^t) \begin{pmatrix} \mathbf{I} & \mathbf{X} \\ \mathbf{X}^t & \mathbf{Y} \end{pmatrix} \begin{pmatrix} \boldsymbol{\omega} \\ \dot{q} \end{pmatrix} \tag{3.1}$$

where t denotes transpose. In this equation, for a nonlinear molecule of N atoms, $\boldsymbol{\omega}$ is a 3-dimensional column vector of the angular-velocity components of the molecule-fixed coordinate system relative to a system whose orientation is fixed in the laboratory. The time derivatives of the vibrational coordinates form the 3N-6 dimensional column vector $\dot{q}$. $\mathbf{I}$ is the 3 x 3 dimensional instantaneous inertial tensor,

$$\mathbf{I} = \begin{pmatrix} I_{xx} & -I_{xy} & -I_{xz} \\ -I_{yx} & I_{yy} & -I_{yz} \\ -I_{zx} & -I_{zy} & I_{zz} \end{pmatrix} \tag{3.2}$$

$$I_{kk} = \sum_{\alpha=1}^{N} m_\alpha (\mathbf{r}_\alpha \cdot \mathbf{r}_\alpha - r_{\alpha k}^2); \quad k = x, y \text{ or } z \tag{3.3a}$$

$$I_{kk'} = \sum_{\alpha=1}^{N} m_\alpha r_{\alpha k} r_{\alpha k'}; \quad k \neq k' \tag{3.3b}$$

where

m_α = mass of the α'th atom
$\mathbf{r}_\alpha$ is the coordinate vector of the α'th atom in the center-of-mass system
$r_{\alpha k}$ denotes the k'th component of the α'th vector.

$\mathbf{Y}$ in Eq. (3.1) is (3N-6) x (3N-6) with elements defined by

$$Y_{ij} = \sum_{\alpha=1}^{N} m_\alpha \left(\frac{\partial \mathbf{r}_\alpha}{\partial q_i}\right) \cdot \left(\frac{\partial \mathbf{r}_\alpha}{\partial q_j}\right) \tag{3.4}$$

$\mathbf{X}$ is 3 x (3N-6) with elements defined by

$$X_{ki} = \sum_{\alpha=1}^{N} m_\alpha \left[\mathbf{r}_\alpha \times \left(\frac{\partial \mathbf{r}_\alpha}{\partial q_i}\right)\right]_k \tag{3.5}$$

where k denotes the k'th component of the cross product. Expanding Eq. (3.1) we obtain

$$2T = \sum_k I_{kk}\omega_k^2 - \sum_{k,k'} I_{kk'}\omega_k\omega_{k'} + \sum_i \sum_j Y_{ij}\dot{q}_i\dot{q}_j + 2\sum_k \sum_i X_{ik}\omega_k \dot{q}_i \tag{3.6}$$

where k,k' = x, y, or z; and i, j run over the vibrational coordinates.

The first two terms are pure rotational terms, the third a pure vibrational term, and the last a Coriolis term. The partitioning of the energy between the pure rotational terms and the Coriolis term depends on the choice of the rotating axis system used to describe the problem[18–22].

In order to obtain a quantum mechanical kinetic energy operator, a momentum representation of the kinetic energy is required. If coordinates are chosen having conjugate momenta

$$\mathbf{P} = \left(\frac{\partial T}{\partial \boldsymbol{\omega}}\right) \tag{3.7}$$

and

$$\mathbf{p} = \left(\frac{\partial T}{\partial \dot{q}}\right) \tag{3.8}$$

where **P** is a 3-dimensional vector of angular momentum components and **p** is a 3N-6-dimensional vector of the momenta conjugate to q, the momentum representation of the kinetic energy is obtained simply by inverting the 3 N-3 x 3 N-3 matrix appearing in Eq. (3.1):

$$2T = (\mathbf{P}^t, \mathbf{p}^t) \begin{pmatrix} \mathbf{I} & \mathbf{X} \\ \mathbf{X}^t & \mathbf{Y} \end{pmatrix}^{-1} \begin{pmatrix} \mathbf{P} \\ \mathbf{p} \end{pmatrix} \tag{3.9}$$

The 3 N-3 x 3 N-3 matrix in Eq. (3.9) is hereafter referred to as the rotation-vibration **G** matrix. Transformation to the correct quantum mechanical from according to Kemble[23] yields

$$2T = g^{1/4} \sum_k \sum_{k'} P_k g^{-1/2} g_{kk'} P_{k'} g^{1/4} + g^{1/4} \sum_i \sum_j p_i g^{-1/2} g_{ij} p_j g^{1/4} + g^{1/4} \sum_k \sum_i \left(P_k g^{-1/2} g_{ki} p_i + p_i g^{-1/2} g_{ik} P_k\right) g^{1/4} \tag{3.10}$$

where $k,k' = x, y$ or z and i, j run over the vibrational coordinates. In this equation, g_{mn} denotes the appropriate element of the rotation-vibration **G** matrix [Eq. (3.9)] and g is the value of its determinant. The P_k are the components of the quantum mechanical angular momenta and the p_i the vibrational momentum operators $\frac{\hbar}{i}\frac{\partial}{\partial q_i}$.
In general, g and the g_{mn} may be functions of the coordinates and thus may not commute with the momenta. The three terms are identified, as before [Eq. (3.6)], as the kinetic energy of pure rotation, pure vibration and vibration-rotation interaction, respectively.

In principle, it is possible to choose a rotating axis system and evaluate numerically the coordinates and coordinate derivatives required to compute the elements of **I**, **X** and **Y** [Eq. (3.1)] for a given dynamical model. By doing this for a grid of values of the 3N-6 vibrational coordinates and obtaining the vibration-rotation **G** matrix by inversion of the $\mathbf{G}^{-1}$ matrix at each grid point, we may obtain a rather complete description of the vibration-rotation kinetic energy [Eq. (3.9)]. By expressing the g_{mn}, $g^{1/4}$ and $g^{-1/2}$ in multidimensional Taylor series or mixed Taylor-Fourier series, an accurate quantum mechanical kinetic energy operator could be written for the dynamical model used and for the particular choice of rotating axes. As pointed out by Gwinn and Gaylord[19)], the solutions of the eigenvalue problems associated with a vibration-rotation problem do not and *must not* depend on the choice of the rotating axis system as long as an adequate Hamiltonian is used. What do depend on the axis system used are the numerical values of elements of the inertial tensor or vibration-rotation interaction constants determined from analysing the data.

For all but the smallest molecules, the procedure outlined in the previous paragraph is impractical. If we can locate or approximate the vibrational band origins from the experimental data, simplifications result. We then treat the J = 0 states in which case all terms in Eq. (3.1) involving rotational angular momenta vanish. Rewriting the pure vibrational term we obtain the vibrational kinetic energy T_v:

$$2T_v = \sum_i \sum_j p_i g_{ij} p_j + 2V'(q) \tag{3.11}$$

The term $V'(q)$ has been referred to as a "pseudopotential" term because it lends itself to expansion in a Taylor (or Fourier) series in the vibrational coordinates and may be absorbed into the effective potential. $V'(q)$ is given by

$$\begin{aligned} V'(q) &= \frac{\hbar^2}{8} \sum_i \sum_j \left(\frac{\partial}{\partial q_i} g_{ij} \left(\frac{\partial \ln g}{\partial q_j} \right) \right) \\ &+ \frac{\hbar^2}{16} \sum_i \sum_j \left(g_{ij} \left(\frac{\partial \ln g}{\partial q_i} \right) \left(\frac{\partial \ln g}{\partial q_j} \right) \right) \end{aligned} \tag{3.12}$$

The total vibrational Hamiltonian is then given by

$$H_v = -\frac{\hbar^2}{2} \sum_i \sum_j \frac{\partial}{\partial q_i} g_{ij} \frac{\partial}{\partial q_j} + V'(q) + V(q) \tag{3.13}$$

where V (q) is an appropriate potential energy function in the 3N-6 vibrational coordinates.

At this point we introduce approximations to obtain a Hamiltonian appropriate for treating the large amplitude modes. We first consider the high-frequency small-amplitude vibrations. Several simplifications occur as a direct result of the small vibrational amplitude. First, the potential energy expansion may be truncated after the harmonic terms without serious error. Secondly, the **G** matrix elements corresponding to the small-amplitude modes may be taken to be constant and thus commute with the momenta. Thirdly, the pseudopotential terms corresponding to these coordinates may be neglected or simply considered as making a very small contribution to the effective force constants.

There are several possible approaches to simplifying the vibrational Hamiltonian given by Eq. (3.13). Some of these will be outlined here. The first approach consists of removing the harmonic potential energy cross terms and the kinetic energy terms by the equivalent of a normal coordinate transformation. We define a new set of coordinates Q

$$\mathbf{q} = \mathbf{L}\,\mathbf{Q} \tag{3.14}$$

such that the resulting vibrational Hamiltonian is given by

$$H_v = -\frac{\hbar^2}{2}\sum_{i=1}^{3N-6}\frac{\partial^2}{\partial Q_i^2} + V_{eff}(\mathbf{Q}) \tag{3.15}$$

V_{eff} (Q) is the effective potential function including contributions from the pseudopotential [V′(q) in Eq. (3.13)] as well as contributions from the kinetic energy due to the fact that the transformation (3.14) is nonlinear. The effective potential has no cross terms of degree less than cubic by the definition of the transformation **L**. In general, the elements of L are functions of the 3N-6 vibrational coordinates because the g_{ij} [Eq. (3.13)] are functions of the coordinates. However, in some cases, the g_{ij} are rigorously constant, while for other small amplitude coordinates they are effectively constant over the range of values accessible to the coordinate. Consequently, the elements of **L** are generally functions of the large-amplitude coordinates only.

Once Eq. (3.15) has been reached, it is generally possible to separate the small-amplitude (harmonic) modes from the large-amplitude modes and treat any remaining anharmonic coupling terms by perturbation methods. The resulting small-amplitude vibrational Hamiltonian is given by:

$$H_{S.A.} = \frac{1}{2}\sum_{j=1}^{3N-M-6} -\hbar^2\frac{\partial^2}{\partial Q_j^2} + \lambda_j Q_j^2 \tag{3.16}$$

where $\lambda_j = 4\pi^2\nu_j^2$ and M is the number of large-amplitude modes. The λ_j's contain contributions from anharmonic interactions with the large-amplitude modes and diagonal terms higher than harmonic have been neglected.

The large-amplitude Hamiltonian is then given by:

$$H_{L.A.} = -\frac{\hbar^2}{2}\sum_{m=1}^{M}\frac{\partial^2}{\partial Q_m^2} + V_{eff}(Q_m) \tag{3.17}$$

The effective potential contains contributions from anharmonic coupling terms with the small-amplitude coordinates which may be absorbed into the effective potential constants for the large-amplitude vibrational coordiantes. Equation (3.17) or some variation thereof has commonly been used to treat the data for large-amplitude modes. It has the distinct advantage that precise *a priori* knowledge of the dynamical path of the large-amplitude modes is not required to treat the spectral data. It has the disadvantage that the physical meaning of the coordinates is somewhat obscured by the procedure which has been used to derive them.

A number of Hamiltonians related to Eq. (3.17) by linear transformations have been used. For example, coordinates which are not mass weighted so that an effective mass appears explicitly have been used, as have a number of reduced or dimensionless coordinates. The relationships among various coordinates have been discussed by Laane[24)] and by Gibson and Harris[15)].

Another approach to separation of the large- and small-amplitude modes is applicable when the kinetic and potential energy coupling terms between these modes are small. In such cases, a Van Vleck transformation may be used[23)]. The effective kinetic energy operator for the large-amplitude modes then becomes

$$T_{eff} = -\frac{\hbar^2}{2}\sum_{m}^{M}\sum_{l}^{M}\frac{\partial}{\partial q_m}\left[g_{ml} + 2\,\delta_{ml}\sum_{i=1}^{3N-M-6} g_{m_i}^2 \sum_{n_i'}{}' \frac{\langle n_i \left| \frac{\partial}{\partial q_i} \right| n_i' \rangle^2}{E_{n_i} - E_{n_i'}}\right]\frac{\partial}{\partial q_l} \tag{3.18}$$

where some small pseudopotential terms have been neglected. In this equation, m and l run over the large-amplitude coordinates and i over the small-amplitude coordinates. δ_{ml} is the Kronecker delta and n_i is a quantum number for the i'th small-amplitude mode. It should be noted that the g_{ml} and g_{m_i} may be functions of the large-amplitude coordinates.

Similarly, the potential energy cross terms may be removed. If we write the total vibrational potential energy

$$V(q) = V_s(q_s) + V_L(q_L) + \sum_{\substack{\text{cross}\\ \text{terms}}} f_s(q_s)\, f_L(q_L) \tag{3.19}$$

where q_s denotes the manifold of small-amplitude coordinates and q_L the large-amplitude coordinates, then a second order Van Vleck transformation yields an effective large-amplitude potential given by

$$V_{eff}(q_L) = V_L(q_L) + \sum_{\substack{\text{cross}\\ \text{terms}}} \left\{ \langle n_s | f_s(q_s) | n_s \rangle f_L(q_L) + \sum_{n_s'}{}' \frac{\langle n_s | f_s(q_s) | n_{s'} \rangle^2}{E_{n_s} - E_{n_s'}} f_L^2(q_L) \right\} \tag{3.20}$$

Normally, the $f_s(q_s)$ need not be carried past harmonic terms due to the small amplitudes of vibration. It is seen that the small-amplitude coordinates contribute to the effective potential constants in the large-amplitude coordinates.

The effective kinetic and potential energy functions, Eq. (3.18) and (3.20), may still have cross terms between two or more large-amplitude modes. The kinetic energy and harmonic potential energy cross terms may be removed by the equivalent of the nonlinear transformation given in Eq. (3.14) over only the large-amplitude modes leading to the same large-amplitude Hamiltonian as in Eq. (3.17). Alternatively, if the cross terms are small enough, the Van Vleck perturbation formulas may be used to remove cross terms between large-amplitude coordinates. Appropriate choices of large-amplitude coordinates, e. g., polar coordinates for pseudorotation, may facilitate this separation of variables.

These procedures are rather cumbersome, but yield a mechanism by which the dependence of potential functions of large-amplitude modes on the vibrational quantum numbers of small-amplitude modes may be rationalized. They also furnish a procedure by which the potential functions may be extrapolated to a "vibrationless" state if sufficient data on the vibrational dependence of the potential functions are obtained.

The last procedure to be considered is the simplest. This is analogous to the separation of high and low frequencies given by Wilson, Decius and Cross[25] for small-amplitude vibrations. In this procedure, cross terms in the potential energy between modes of large and small amplitudes are simply neglected. In order to obtain a kinetic energy operator for the large-amplitude modes, only these modes are included in calculating the vibration-rotation inverse **G** matrix, Eq. (3.1). If we wish to use the Hamiltonian given by Eq. (3.13) in that form, it is then necessary to determine the g_{ij} and pseudopotential as functions of the large-amplitude coordinates. What is required is to calculate the coordinate vectors and their derivatives for each atom in a center-of-mass system for a grid of values of the large-amplitude coordinates. At each grid point, Eqs. (3.2) to (3.5) are used to calculate the $\mathbf{G}^{-1}$ matrix, and the **G** matrix Eq. (3.9) is obtained by inverting the G^{-1} matrix. The determinant of **G** as well as the appropriate **G** matrix elements may be expressed as Taylor or mixed Fourier/Taylor series and used to form the Hamiltonian Eq. (3.13).

The above calculations require assumption of a dynamical model for the large-amplitude modes. A much simpler model is to use a constant-reduced-mass Hamiltonian similar to that given by Eq. (3.17). This is by far the most common procedure. It may be justified by considering a molecule with a single large-amplitude mode such as a ring puckering. If x is the ring-puckering coordinate, we may define Q_x by a nonlinear transformation

$$Q_x = \int [g_{xx}(x)]^{-1/2} dx \qquad (3.21)$$

where $g_{xx}(x)$ is the ring-puckering **G** matrix element expressed as a function of the coordinate[18]. This leads directly to the Hamiltonian of Eq. (3.17) for a single large-amplitude mode. While this procedure is the most straight-forward, there is no direct mechanism by which the coupling of large- and small-amplitude vibrational modes may be taken into account. The existence of such coupling is observable in combination bands in the mid infrared or Raman spectrum that involve both high-frequency and large-amplitude modes; excitation of one or more quanta of the former leads to slightly different effective potential functions for the latter. This will be discussed later in more detail.

C. One-Dimensional Hamiltonians

1 Symmetrical Potential Functions

Four-membered ring molecules have one out-of-plane ring vibration, usually referred to as the ring-puckering vibration. In saturated four-membered rings this mode usually has the lowest frequency. It may have a large amplitude and be quite anharmonic. A one-dimensional Hamiltonian may be used if the coupling with the small-amplitude modes may be neglected or absorbed into the effective Hamiltonian. The most common one-dimensional Hamiltonian used to interpret the spectral data for one-dimensional ring puckering with symmetrical potential functions has been

$$H = -\frac{\hbar^2}{2\mu}\frac{d^2}{d\bar{x}^2} + \bar{a}\,\bar{x}^4 + \bar{b}\,\bar{x}^2 \tag{3.22}$$

where $\bar{x}$ is a ring-puckering coordinate, the bar indicating that the reduced mass is assumed to be constant, and μ is the associated reduced mass evaluated for a infinitesimal displacement from the planar conformation. Figure 3.1 indicates one possible definition of the ring-puckering coordinate $\bar{x}$, for cyclobutane, as half the perpendicular distance between ring diagonals. Implicit in the use of Eq. (3.22) is the nonlinear transformation that removes the coordinate dependence of the reduced mass and absorbs the pseudopotential terms into the effective potential constants, as well as any zero-point vibrational averaging over the other modes of the molecule.

The potential function in Eq. (3.22) is remarkably simple. Nevertheless, it is applicable to the description of a variety of molecular systems. For a = 0, b > 0 it represents a harmonic oscillator; for a > 0, b = 0 a quartic oscillator; for a > 0, b > 0 a single mini-

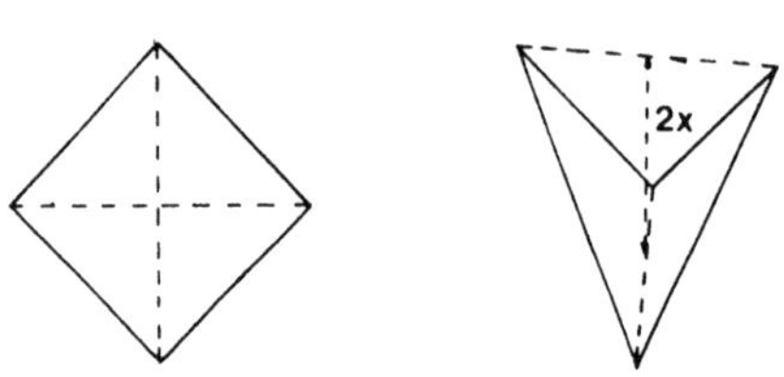

Fig. 3.1. One possible definition of a ring-puckering coordinate for four-membered ring molecules as half the perpendicular distance between the ring diagonals.
[Reproduced from Malloy, T. B., Jr., Lafferty, W. J.: J. Mol. Spectroscopy *54*, 20 (1975).]

mum quartic-quadratic oscillator; for $a > 0$, $b < 0$ a double minimum oscillator with a barrier height given by $b^2/4a$.

Except for the obvious case of the one-dimensional harmonic oscillator ($a = 0$), the solutions to the Schrödinger equation corresponding to the Hamiltonian of Eq. (3.22) cannot be given in closed form. Several numerical techniques have been used, including application of the linear variation method with a truncated harmonic oscillator basis set and various numerical integration techniques[26)]. There are advantages and disadvantages for each of the techniques used. While the numerical solution of the Schrödinger equation for ring-puckering problems was not a routine matter fifteen to twenty years ago, with today's digital computers it has become so. The relative speeds of the various techniques now result in differences of a few seconds or fractions of seconds of computer time in the treatment of one-dimensional Hamiltonians.

Historically, the most common technique used has been the linear variation method. In this procedure, the wave functions are expressed as linear combinations of harmonic-oscillator basis functions

$$\psi_m = \sum_{i=0}^{n} t_{im}\phi_i \tag{3.23}$$

where ψ_m is the wave function corresponding to E_m, the m'th level. The ϕ_i are appropriate Hermite functions and n represents the highest quantum number attained before truncation. Since this is an orthonormal basis, application of the linear variation method corresponds to finding the eigenvalues of the Hamiltonian matrix, with elements defined by

$$H_{jk} = \int \phi_j H \phi_k \, d\bar{x} \tag{3.24}$$

where H is the operator given in Eq. (3.22). Formally, the variation approximation to the energy eigenvalues is given by

$$\Lambda_E = \mathbf{T}^t \mathbf{H}\,\mathbf{T} \tag{3.25}$$

where the elements of **T** appear in Eq. (3.23). Considerable reduction in the computer time may be achieved by factoring the matrix into even and odd blocks which reduces the computer time by approximately a factor of four. The basis set may then be represented by

$$\psi_m^{even} = \sum_{i=0}^{n\ even} t_{im}\phi_i^{even} \tag{3.26a}$$

$$\psi_{m'}^{odd} = \sum_{i'=1}^{n\ odd} t_{i'm'}\phi_{i'}^{odd} \tag{3.26b}$$

In addition, the number of basis functions required to obtain a satisfactory representation depends on the choice of the harmonic frequency for the basis. Stated in another way, the harmonic oscillator basis functions used should have appreciable amplitude over a range of values of the coordinate corresponding to that for which the desired wave functions ψ_m have appreciable amplitude[27].

It is obvious that if the harmonic frequency chosen is too high, so that the classical turning points for the highest functions used are inside the classical turning points of the highest level desired accurately, it will not be possible to synthesize a satisfactory wave function for this level. Correspondingly, if the harmonic frequency is too low, a larger number of basis functions will be required to eliminate unwanted amplitude at values of the coordinate outside the classical turning points. Reid[28] pointed out that problems of this nature had caused truncation errors in the treatment[29] of the quartic oscillator some years earlier. The discrepancies were not serious in the eigenvalues, but, as might be expected, were more noticeable in the values of the matrix elements of the operators x^2 and x^4. As pointed out by Reid, the convergence of the expansions, Eqs. (3.26a, b) is a slowly varying function of the harmonic scale factor and it is probably not worth treating this as a variation parameter. Carreira, Mills and Person[30] have given a rule of thumb by which they choose the harmonic frequency so that the classical turning points of the basis set for the highest level used correspond to the classical turning points at approximately twice the energy of the last level for which an accurate eigenvalue is desired.

Typically, the above considerations are less compelling for one-dimensional problems since the choice of a less than optimum scale factor can be compensated by simply using more functions. The number of basis functions required depends on the number of levels for which accurate eigenvalues and eigenvectors are desired and on the nature of the potential function. Usually the number of eigenvalues required lies between about 10 and 20. If the potential function is a single-minimum function and a judicious choice of scale factor is made, 40 basis functions (20/symmetry block) are more than sufficient. On the other hand, the number required for a double-minimum potential function depends on the barrier height, a larger number being required for a higher barrier. For barriers up to 2–3 kcal/mole in ring-puckering problems, between 50 and 70 basis functions (25–35/block) have proved to be sufficient. One exception to this arises when the spacing between the levels of an inversion doublet below the barrier is required to a high degree of accuracy.

Most of the variation calculations done up to about 1965 used the Jacobi diagonalization method[31] for finding the eigenvalues and eigenvectors. In this method, if 70 basis functions were used, all 70 eigenvalues and eigenvectors were found and then arranged in order of increasing energy. More recently, in numerous applications, the Givens-Householder method[31] has been used. Since it is possible to generate the eigenvalues in increasing order, only those eigenvalues which are desired are found, their eigenvectors generated and the process terminated. Since 10–20 eigenvalues, rather than 60–70 are typically generated, a significant reduction in computer time is realized. This is over and above the fact that the time required for the Givens-Householder diagonalization is generally less than that of the Jacobi technique even when all the eigenvalues and eigenvectors are generated by both methods. The efficiency of this method of calculation, while perhaps less than that of the Numerov-Cooley numerical integration, is quite good[26]. Typically, fitting vibrational data up to the 15'th excited state, performing three cycles of a least squares iteration and calculating intensities etc.,

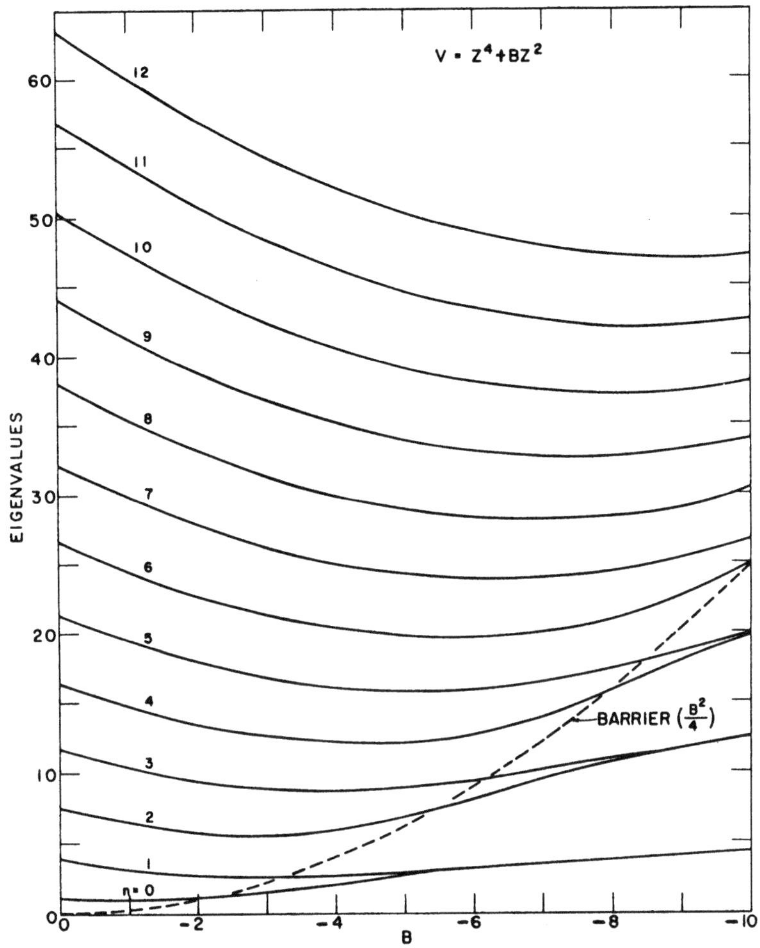

Fig. 3.2. Variation of the eigenvalues for the potential function $V(Z) = Z^4 + BZ^2$ *vs.* B. [Reproduced from Laane, J., Lord, R. C.: J. Chem. Phys. *47*, 4941 (1967).]

with 70 basis functions required *ca.* 8 seconds of computer time on a Univac 1108 computer.

One set of reduced coordinates used leads to a Schrödinger equation

$$-A \frac{d^2\psi}{dZ^2} + A(Z^4 + BZ^2)\psi = A\lambda\psi \qquad (3.27)$$

where $x = (\hbar^2/2\mu\bar{a})^{1/6} Z$ and $E = A\lambda$. Figure 3.2 shows the variation of the eigenvalues λ with B for $B \leq 0$. The limit as B becomes large and positive would be a harmonic oscillator. As B becomes more negative, the levels below the reduced barrier (given by $B^2/4$) coalesce into pairs of inversion doublets as shown in Fig. 3.2. Laane[24)] has published tables of eigenvalues for a grid of values of B. These have been very useful for making assignments and obtaining initial estimates of the potential constants from ring-puckering vibrational data.

2 Reduced-Mass Calculations

In the interpretation of the spectral data it is usually the constants A and B in Eq. (3.27) or some other set of reduced potential constants that are evaluated. The barrier height, $AB^2/4$ for $B < 0$, is thus determined directly. However, if one wants to relate the value of the dimensionless parameter $|Z|$ at the minimum of a double-minimum potential function to the absolute geometry of the puckered ring, the reduced mass must be known in order to find $\bar{x}$ from $|Z|$. One is then required to introduce assumptions about the dynamical model of the ring-puckering motion.

For cyclobutane, the motion of the four carbon atoms is relatively unambiguous. We first assume we can neglect bond stretching as the molecule puckers. Secondly, we require that the non-bonded C···C distances change symmetrically. For a molecule like trimethylene oxide, the choice is not so clear. Gwinn and co-workers[32] have defined a parameter ω, which expresses the relative amount of bending about the C···0 diagonal and the C···C diagonal. First a parameter ρ is defined

$$\rho = \Delta(C\cdots 0)^2/\Delta(C\cdots C)^2 \tag{3.28}$$

where $\Delta(C\cdots 0)^2$ and $\Delta(C\cdots C)^2$ are the respective changes in the squares of the non-bonded C···0 distance and the non-bonded C···C distance. Then ω is defined as

$$\omega = (1 - \rho)/(1 + \rho) \tag{3.29}$$

and varies between +1 and −1. The value $\omega = +1$ corresponds to the ring bending about the C···0 diagonal along while the value $\omega = -1$ corresponds to bending about the C···C diagonal. The value $\omega = 0$ corresponds to symmetrical bending about each diagonal, as in cyclobutane.

Model calculations to reproduce the variation of rotational constants with vibrational state are sensitive to the value of ω. Thus an approximate value may be determined from them that yields experimental information about the dynamics of the vibration. If the Hamiltonian with variable reduced mass is used, the dependence of the reduced mass on coordinate will be a function of ω. On the other hand, if the constant effective mass model is used, the reduced mass can be evaluated from Eqs. (3.1) to (3.10) for an infinitesimal displacement from the planar conformation. The constant effective reduced mass derived in this fashion is independent of ω, and thus, no knowledge of ω is needed to use the Hamiltonian Eq. (3.22); conversely, no information about the value of ω can be determined from the vibrational data and Eq. (3.22).

Probably the most serious source of uncertainty in calculating a reduced mass concerns the details of the motion of CH_2 groups during the puckering vibration. Usually, there will be one or more CH_2 rocking modes, only a factor of three or so higher in frequency than, and of the same symmetry species as, the ring puckering. This means that quadratic kinetic energy and potential energy interaction terms will enter into the Hamiltonian. These terms, more than any others, lead to different forms for the ring-puckering coordinate for isotopic species. The details of the motion of CH_2 groups during the ring-puckering vibration can have a large effect on the reduced mass for this motion. A bisector model, i. e., one in which the H-C-H angle remains constant and shares a common bisector with the adjacent ring angle seems to work reasonably well for trimethylene oxide. The quartic terms in the potential are quite reasonably predicted for various isotopic species by this model as discussed below for trimethylene oxide. On the other hand, Stone and Mills[33] found that the reduced-mass ratio calculated for cyclobutane-$\underline{d}_8$ and cyclobutane for a bisector model, 1.549, underestimated the isotopic shift. They empirically adjusted this ratio to 1.641 to obtain better agreement with the experimental results. Malloy and Lafferty have discussed this point in some detail[27].

To summarize, uncertainties in the form of the puckering coordinate can cause uncertainties in the calculated reduced masses which may lead to errors of several degrees in determination of the puckering angle corresponding to the minimum. The use of a variable-reduced-mass Hamiltonian [Eq. (3.14)] rather than the constant-reduced-mass Hamiltonian [Eq. (3.15)] has a minor effect. If the same dynamical model is used, differences in the calculated angles are generally less than a degree. Malloy and Lafferty[27] also considered the effect of errors in the structural parameters and found them to be quite negligible, assuming the errors were not more than ∿ 5° in a bond angle or 0.03 – 0.04 Å in a bond distance.

3 Asymmetric One-Dimensional Hamiltonians

In the previous sections, we have been considering ring molecules for which the odd-power terms in the ring-puckering potential functions must vanish by symmetry. For molecules like mono-substituted cyclobutane derivatives, symmetry no longer dictates that these terms are zero. Consequently, through fourth degree, the constant-effective-mass Hamiltonian analogous to Eq. (3.25) is given by

$$H = -\frac{\hbar^2}{2\hat{\mu}}\frac{d^2}{d\hat{x}^2} + \hat{a}\hat{x}^4 + \hat{b}\hat{x}^2 + \hat{c}\hat{x}^3 + \hat{d}\hat{x} \qquad (3.30)$$

where $\hat{\mu}$ is the reduced mass evaluated for infinitesimal displacement from the coordinate zero, taken as the planar ring conformation. However, a Hamiltonian which represents the same system and has fewer adjustable parameters is obtained by translating the origin by an amount δ to correspond to an extremum of the potential function, that is, by the transformation $\bar{x} = \hat{x} + \delta$:

$$H = -\frac{\hbar^2}{2\mu}\frac{d^2}{dx^2} + \bar{a}\bar{x}^4 + \bar{b}\bar{x}^2 + \bar{c}\bar{x}^3 \qquad (3.31)$$

The dimensionless analog of this equation [see Eq. (3.27)] is

$$H = A\left(-\frac{d^2}{dz^2} + Z^4 + BZ^2 + CZ^3\right) \qquad (3.32)$$

In fitting vibrational energy separations, Eq. (3.32) or an equivalent reduced equation is used.

The calculation of the equilibrium value of the dihedral angle for a puckered unsymmetrical ring from the potential function is not straight forward, as it is in the symmetrical case. Because of the lack of symmetry there is no way *a priori* of determining the conformation at the point of minimum energy ($Z = 0$), or the value of Z for the planar ring. Thus in the treatment of asymmetric ring molecules, the roles played by vibrational spectroscopy and microwave spectroscopy are truly complementary. Since the unscaled potential function in Eq. (3.32) has two adjustable parameters and since odd power terms are no longer excluded from the microwave rotational constant expansions, there are too many parameters to be specified by microwave data alone. Consequently, determination of the potential energy function in reduced coordinates has customarily been done by fitting transitions observed in the far-infrared or Raman spectra or both. On the other hand, the microwave data must be used to characterize the conformation or conformations corresponding to the minimum or minima in the potential function. The structure of the lowest energy conformer, if determined, may then be used as the reference point in the calculation of the reduced mass.

For purposes of comparison, it is possible to classify the various types of potential functions which may be represented by the functional form used in Eq. (3.32) with a few simple considerations. The restrictions we shall make are always to locate the origin in the minimum, or if more than one, in the deepest minimum; second minima or inflection points are restricted to negative values of the coordinate Z; and the positive values of Z always represent the most rapidly rising portion of the function. These restrictions do not eliminate any unique shape of potential function. Any other functions described by Eq. (3.32) are related to those already included by a simple translation of the origin or by rotation about the vertical axis. These operations, at most, change the eigenvalues by an additive constant. The different types of potential functions are summarized in Table 3.1.

It is a simple matter to restrict the potential functions as mentioned above. First, the parameter A [Eq. (3.32)] is simply a scale factor and need not be considered further. If both B and C, the respective coefficients of Z^2 and Z^3, are restricted to positive values, the origin will be a minimum and second minima or inflection points, if present, will occur for negative values of Z. If we make the further restriction that $9C^2 < 36B$, the origin will be in the deepest minimum. The case $9C^2 = 36B$ represents a symmetric double minimum potential function with the origin in the right well.

With the above in mind, all asymmetric double-minimum potential functions represented by Eq. (3.32) may be described with parameters in the range $36B > 9C^2 > 32B$. The maximum occurs at

$$Z = (-3C + \sqrt{9C^2 - 32B})/8$$

and the second minimum at

$$Z = (-3C - \sqrt{9C^2 - 32B})/8.$$

For $9C^2$ nearer the limit of $36B$, the second minimum is well below the barrier separating the two minima. The squared wave functions of the states below the

Table 3.1. Classification of asymmetric potential functions. ($Z^4 + BZ^2 + CZ^3$); $36B \geq 9C^2 \geq 0$

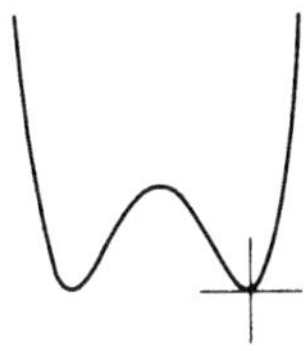	$9C^2 = 36B$ Symmetric double minimum potential with the origin in the right well
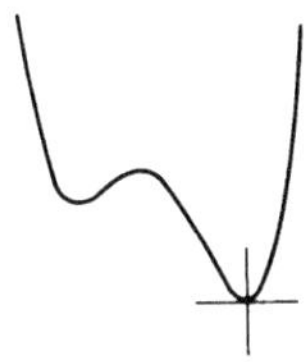	$36B > 9C^2 > 32B$ Asymmetric double minimum potential function with the origin in the deeper minimum on the right. The maximum occurs at $Z = (-3C + \sqrt{9C^2 - 32B})/8$ and the shallower second minimum at $Z = (-3C - \sqrt{9C^2 - 32B})/8$
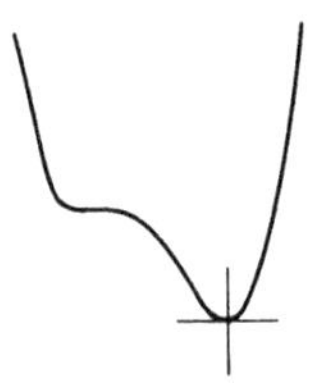	$9C^2 = 32B$ Asymmetric single minimum potential function with an inflection point with a horizontal slope. This inflection point occurs at $Z = -3C/8$
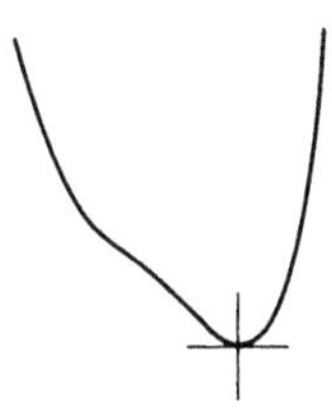	$32B > 9C^2 > 24B$ Asymmetric single minimum potential function with two inflection points. These inflection points occur at $Z = (-3C \pm \sqrt{9C^2 - 24B})/12$.
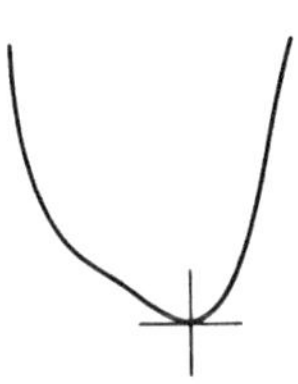	$9C^2 = 24B$ Asymmetric single minimum potential function with one inflection point at $Z = -C/4$
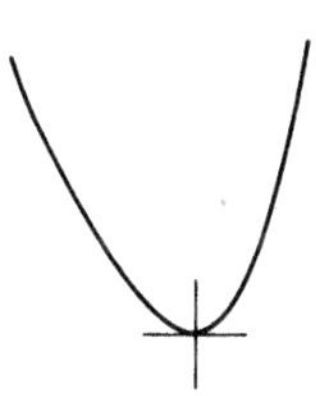	$24B > 9C^2 > 0$ Asymmetric single minimum potential function with no inflection points
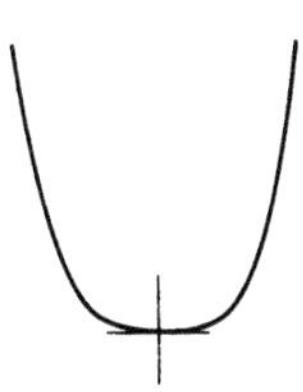	$9C^2 = 0$ Symmetric single minimum quartic-quadratic potential function

barrier show a definite "left well" or "right well" character for the states. For any state above the barrier, the probability density has maxima across the full range of Z accessible to that state. Vibrational transitions below the barrier follow left well well ⟷ left well or right well ⟷ right well selection rules for both far-infrared and Raman transitions. Above the barrier, these identifications are more difficult to make. Even very slight asymmetry drastically reduces the probability of tunneling compared to the case of the symmetric double-minimum function.

As $9C^2 \to 32B$, the second minimum becomes less pronounced. At $9C^2 = 32B$, instead of a maximum and a second minimum, there is an inflection point with a horizontal slope at $Z = -3C/8$. For $32B > 9C^2 > 24B$, there is a single minimum with inflection points in the potential function at

$$Z = (-3C \pm \sqrt{9C^2 - 24B})/12.$$

The vibrational spectra for potential functions of this type are characterized by negative anharmonicity for the first few transitions, with the frequency eventually reaching a minimum, and positive anharmonicity for the higher transitions. The quadratic term in the potential is primarily responsible for the value of the 0–1 frequency. For small amplitudes, i.e. for the lower transitions, the cubic term is dominant in determining the anharmonicity which in this case is negative. At larger amplitudes, i. e. higher energy, the quartic term balances the effect of the cubic term, causing the transitions to reach a minimum frequency. The positive anharmonicity due to the quartic term then dominates for the higher transitions. These characteristics may be observed for potential functions of this type unless the Boltzmann factor intervenes to depopulate levels above the inflection points.

The last general category of asymmetric potential functions is that for which $24B > 9C^2 \geq 0$. For $24B = 9C^2$, there is a single inflection point at $Z = -C/4$. For functions in the range described, there are no inflection points. For functions with $9C^2$ near $24B$, the characteristics mentioned in the previous paragraph are applicable, with the transition of minimum frequency being reached more rapidly, followed by positive anharmonicity due to the quartic term. Eventually, the effect of the cubic term becomes such that the anharmonicity is immediately positive. The limit as $9C^2 \to 0$ is, of course, the symmetric single-minimum quartic-quadratic oscillator considered earlier.

The above classification of asymmetric potential functions is convenient for comparison of different molecules or as a systematic basis for making an initial fit to experimental data. However, when the Schrödinger equation is being solved by the linear variation method with harmonic-oscillator basis functions, it may not provide the best choice of origin for the basis function. For example, a better choice in the case of an asymmetric double-minimum oscillator, where accurate solutions are required in both wells, would be somewhere between the two wells. Systematic variation of the parameters may still be made as outlined above, but the origin should be translated before the Hamiltonian matrix is set up. The equations given earlier

(3.23) to (3.26) that describe the Hamiltonian matrix in the harmonic basis are valid, with the exception that the matrix does not factor into odd and even blocks. The Z^3 matrix elements connect these two blocks.

For potential functions with a single minimum (or a very shallow second minimum), location of the origin in the minimum is probably the best choice to obtain most rapid convergence of the variation functions. Again, as for the symmetric cases, some care should be exercised in choosing the harmonic scale factor for the basis, to insure that the truncated basis has sufficient flexibility to produce amplitude in the classically allowed regions and to cancel amplitude in the unallowed regions.

D. Two-Dimensional Hamiltonians

Saturated five-membered-ring molecules have two low-frequency out-of-plane ring vibrations and these modes may couple. Equation (3.33) is an appropriate two-

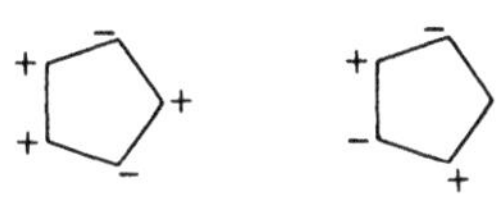

dimensional Hamiltonian with a constant effective mass for molecules where terms of odd powers are excluded by symmetry

$$H = -\frac{\hbar^2}{2\mu_x}\frac{\partial^2}{\partial \bar{x}^2} - \frac{\hbar^2}{2\mu_y}\frac{\partial^2}{\partial \bar{y}^2} + a_1\bar{x}^4 + b_1\bar{x}^2 + a_2\bar{y}^4 + b_2\bar{y}^2 + c_{12}\bar{x}^2\bar{y}^2 \tag{3.33}$$

where $\bar{x}$ is a ring-bending coordiate and $\bar{y}$ is a ring-twisting coordinate. μ_x and μ_y are the associated reduced masses. It is sometimes possible to separate the variables in Eq. (3.33) approximately and thereby to obtain an effective one-dimensional Hamiltonian for the ring-bending vibration. This is possible primarily when one of the modes is of large amplitude while the other is of small amplitude.

1 Pure Pseudorotation

In 1947, Kilpatrick, Pitzer and Spitzer[3)] introduced the notion of pure pseudorotation to explain the thermodynamic data on cyclopentane. In this case, the amplitudes of the ring-bending and ring-twisting coordinates are comparable and Eq. (3.33) as it stands is not even approximately separable. However, transformation to polar coordinates yields a small-amplitude (radial) and a large-amplitude (angular) coordinate. The resulting Schrödinger equation is separable, or approximately so, when pseudorotational barriers are small compared to the barrier to planarity.

Gwinn and co-workers[34)] have given an excellent exposition of the theory appropriate to treating hindered pseudorotation, with particular attention to the use of an angular Hamiltonian for small-barrier cases. For the special case of pure pseudorotation, the potential energy is independent of the angular coordinate. The constant-reduced-mass Hamiltonian [Eq. (3.33)] with restrictions appropriate to pseudorotation, $\mu_1 = \mu_2$, $2\bar{a}_1 = 2\bar{a}_2 = \bar{a}_{12}$, $\bar{b}_1 = \bar{b}_2$, gives the following Schrödinger equation:

$$\left(-\frac{\partial^2}{\partial Z_1^2} - \frac{\partial^2}{\partial Z_2^2}\ 2\right)\psi + (Z_1^2 + Z_2^2)\psi + B(Z_1^2 + Z_2^2)\psi = \lambda\psi \tag{3.34}$$

where

$$Z_1 = (\hbar^2/2\mu\bar{a})^{-1/6}x$$

$$Z_2 = (\hbar^2/2\mu\bar{a})^{-1/6}y$$

$$\lambda = E/A$$

Figure 3.3 gives a potential energy contour diagram appropriate for pure pseudorotation.

In polar coordinates ρ and θ with

$$Z_1 = \rho\cos\theta$$

$$Z_2 = \rho\sin\theta$$

the Schrödinger equation is

$$\left(-\frac{1}{\rho}\frac{\partial}{\partial\rho}\rho\frac{\partial}{\partial\rho} + \frac{1}{\rho^2}\frac{\partial^2}{\partial\theta^2}\right)\psi + (\rho^4 + B\rho^2)\psi = \lambda\psi \tag{3.35}$$

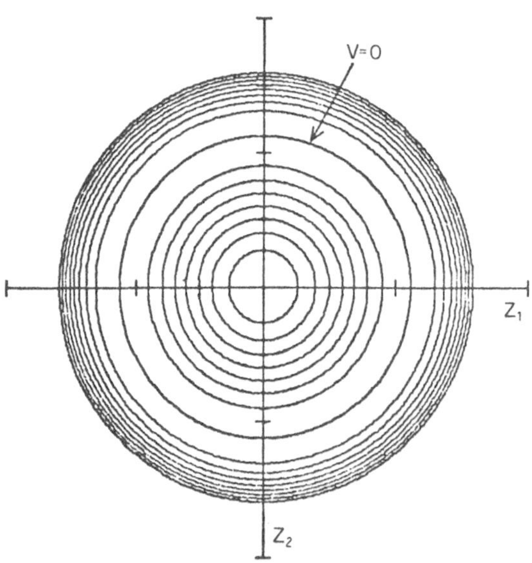

Fig. 3.3. Potential energy contour diagram appropriate to pseudorotation. An energy maximum occurs at the planar conformation, $Z_1 = Z_2 = 0$. The minimum energy track is denoted as V = 0. [Reproduced from Harris, D. O., Engerholm, G. G., Tolman, C. A., Luntz, A. C., Keller, R. A., Kim, H., Gwinn, W. D.: J. Chem. Phys. *50*, 2438 (1969).]

For cyclopentane, the case considered by Pitzer et al.[3], B is large in magnitude and negative. This corresponds to a high barrier to planarity, given by $AB^2/4$, and approximate solutions may be found. For a high barrier to planarity, the value of $1/\rho^2$ may be replaced by its average value and the energy eigenvalues approximated from

$$-\beta \frac{d^2\Theta}{d\theta^2} = E\Theta \tag{3.36}$$

where

$$\beta = \langle 1/\rho^2 \rangle$$

The solutions of Eq. (3.49) are

$$\Theta = (1/2\pi)^{1/2} e^{il\theta}, l = 0, \pm 1, \pm 2, \ldots \tag{3.37}$$

The transition frequencies, $l \rightarrow l + 1$, are given by

$$\nu = A\beta(2l + 1) \tag{3.38}$$

which are equally spaced transitions separated by $2A\beta$.

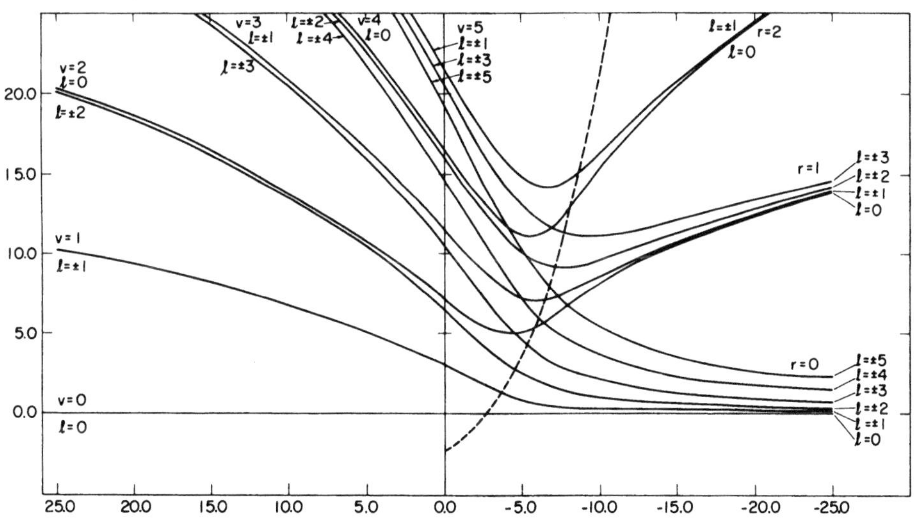

Fig. 3.4. Some of the eigenvalues for the reduced potential $(Z_1^2 + Z_2^2)^2 + B(Z_1^2 + Z_2^2)$ as a function of B [Eq. (3.34)]. The dashed line indicates the top of the barrier. The case of pseudorotation is the limit, on the right, for large negative values of B. Large positive values of B, on the left, correspond, in the limit, to a two-dimensional isotropic harmonic oscillator. [Reproduced from Ikeda, T., Lord, R. C., Malloy, T. B., Ueda, T.: J. Chem. Phys. *56*, 1434 (1972).]

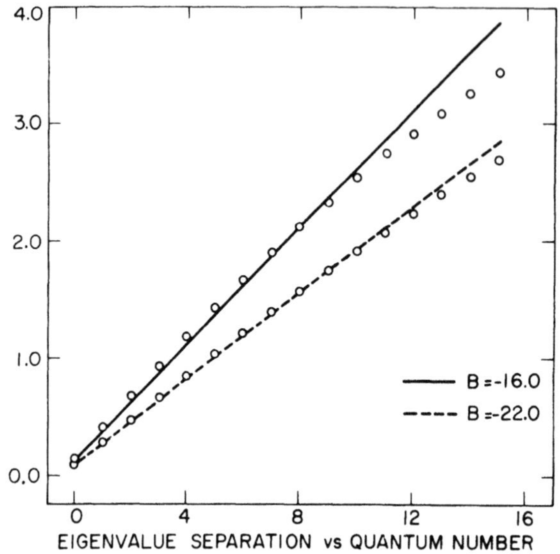

Fig. 3.5. Pseudorotational transitions, $l \rightarrow l+1$ *vs.* l, for the numerical solutions of Eqs. (3.34) and (3.35). The lower the barrier (smaller magnitude of B), the more pronounced the curvature [See Eq. (3.39)].
[Reproduced from Ikeda, T., Lord, R. C., Malloy, T. B., Ueda, T.: J. Chem. Phys. *56*, 1434 (1972).]

Ikeda et al.[35] considered solutions of Eqs. (3.34) and (3.35) by the variation method with two-dimensional harmonic oscillator functions in Cartesian and polar coordinates, respectively, as basis functions. Some of the eigenvalues are plotted in Fig. 3.4 as a function of the parameter B. The case of pure pseudorotation corresponds to large negative values of B on the right hand side of the figure.

It was found from numerical solution of Eqs. (3.34) and (3.35) that the calculated pseudorotational frequencies exhibited curvature rather than a strict linear dependence on quantum number as in Eq. (3.38). This is shown in Fig. 3.5. Such curvature had been experimentally observed for 1,3-dioxolane and tetrahydrofuran[36] and was also noted by Davis and Warsop[37], who used it to estimate the barrier to planarity.

Consideration of approximate solutions to Eq. (3.35) by obtaining an effective Hamiltonian by a 2nd order Van Vleck transformation led to an expression[35] for $l \rightarrow l+1$ transitions given by

$$\nu = A\beta(2l+1) - AD(4l^3 + 6l^2 + 4l + 1) \tag{3.39}$$

with β exhibiting a dependence on the radial quantum number v_ρ given by

$$\beta = -2/B \left[1 - \left[12/(-2B^3)^{1/2}\right] (v_\rho + 1/2)\right] \tag{3.40}$$

and

$$D = 4/B^4 \tag{3.41}$$

The pseudocentrifugal distortion constant, D, accounts for the curvature of the frequencies (Fig. 3.5) and Eq. (3.40) gives the variation of pseudorotational constants with vibrational state.

2 Hindered Pseudorotation

If there is an angular dependence of the potential function, it is still possible to separate the Schrödinger equation approximately in polar coordinates if the angular barriers are much lower than the barrier to planarity. As mentioned earlier, Gwinn et al.[34)] have given an excellent treatment of this case. This has been applied to the interpretation of the microwave and far infrared spectra of tetrahydrofuran and 1,3-dioxolane[36, 38)]. Equation (3.33) may be transformed to mass weighted polar coordinates

$$x = \mu_1^{1/2} r \cos \phi \tag{3.42a}$$

$$y = \mu_2^{1/2} r \sin \phi \tag{3.42b}$$

and expressed as

$$H = -\frac{\hbar^2}{2}\left(\frac{1}{r}\frac{\partial}{\partial r} r \frac{\partial}{\partial r} + \frac{1}{r^2}\frac{\partial^2}{\partial \phi^2}\right) + \mathbf{A} r^4 + \mathbf{B} r^2 + \mathbf{C} r^2 \cos 2\phi + \mathbf{D} r^4 \cos 2\phi + \mathbf{E} r^4 \cos 4\phi \tag{3.43}$$

If the conditions mentioned above are met, the Hamiltonian may be averaged over the radial coordinate, yielding the following Schrödinger equation

$$-B\frac{d^2\psi}{d\phi^2} + \sum_{n=1}^{2} \frac{V_{2n}}{2}(1 - \cos 2n\phi)\psi = E\psi \tag{3.44}$$

where

$$B = \frac{\hbar^2}{2} \langle v_r | \frac{1}{r^2} | v_r \rangle \tag{3.45a}$$

$$V_2 = 2(\mathbf{C} \langle v_r | r^2 | v_r \rangle + \mathbf{D} \langle v_r | r^4 | v_r \rangle) \tag{3.45b}$$

$$V_4 = 2\mathbf{E} \langle v_r | r^4 | v_r \rangle \tag{3.45c}$$

The energy origin has been translated so that $E \geq 0$. Equation (3.44) is of the same form as the Schrödinger equation for a two-fold internal rotor. Higher order terms than four-fold may arise from

(a) higher-degree terms than quartic in the polynomial representation of the potential surface;

(b) a higher degree of approximation than first order in the approximate separation of variables.
The latter will also introduce pseudocentrifugal distortion terms.

The solution of Eq. (3.44) may be accomplished by the variation method, using the free-rotor functions given in Eq. (3.37) as a basis set, and factoring the matrices into even, odd blocks according to $|l|$. On the other hand, additional symmetry factoring is possible by choosing a sine-cosine basis related to the free-rotor basis by a variation of the Wang transformation. Lewis et al.[39] have described a computer program based on this approach.

In some molecules even an approximate separation of variables is not possible. Cyclopentanone is a good example of such a molecule; it is discussed as a special case in Section IV. B.

IV Summary of Results with Examples

A considerable number of large-amplitude vibrations in ring molecules have been treated by the theoretical methods discussed in Section III. Reviews of these studies have been given by Laane[40], Blackwell and Lord[41], Gwinn and Gaylord[19], and Wurrey, Durig and Carreira[42].

We have chosen to illustrate the applications of the theory to specific molecules in the same framework as that of Section III. While our list of applications is not exhaustive, we believe that all of the various aspects of the theoretical treatment are illustrated by at least one example and no important molecules are omitted from the discussion.

A. Molecules Treated by One-Dimensional Hamiltonians

1 Symmetric Systems

a) Four-Membered Ring Molecules

(i) Oxetanone-3 and Thietanone-3. Figure 2.1 shows the far-infrared spectrum of oxetanone-3 obtained with the Jarrell Ash 78–900 vacuum grating spectrophotometer at MIT[10]. The ring puckering in this molecule is nearly harmonic, with slight positive anharmonicity due to the quartic term. This term is of such magnitude that second-order perturbation theory is quite adequate to reproduce the energy-level pattern. The frequencies are given by

$$\nu_{v,\,v+1} = 2AB^{1/2} + (v+1)A/B - 3A(v^2 + 4v + 4)/4B^{3/2} \tag{4.1}$$

Table 4.1 compares the observed frequencies with those calculated by a least squares adjustment of the potential constants using the linear variation method as incorporated in a computer program written by Ueda and Shimanouchi[12]. The potential

Table 4.1. Observed and calculated far infrared transition frequencies for oxetanone-3

Transition	Obs. freq.	Calc. freq.	Obs. – calc.
0–1	140.0	140.13	–0.13
1–2	141.5	141.51	–0.01
2–3	143.0	142.87	0.13
3–4	144.3	144.17	0.12
4–5	145.5	145.46	0.03
5–6	146.6	146.71	–0.11
6–7	147.5	147.93	–0.43

function indicates that the planar ring skeleton of oxetanone-3 corresponds to an energy minimum (Fig. 4.1).

The microwave spectrum of oxetanone-3 has been studied by Gibson and Harris[15]. In the case of small-amplitude harmonic vibrations, the rotational constants should vary linearly with vibrational quantum number. For a single-minimum anharmonic potential representing a large-amplitude coordinate, deviation from this linear dependence is expected on two accounts. If we express the dependence on the large-amplitude coordinate in a power series, it may be necessary to carry the series past the quadratic term. Also, the contribution of the quartic term in the potential energy may cause deviations from linearity.

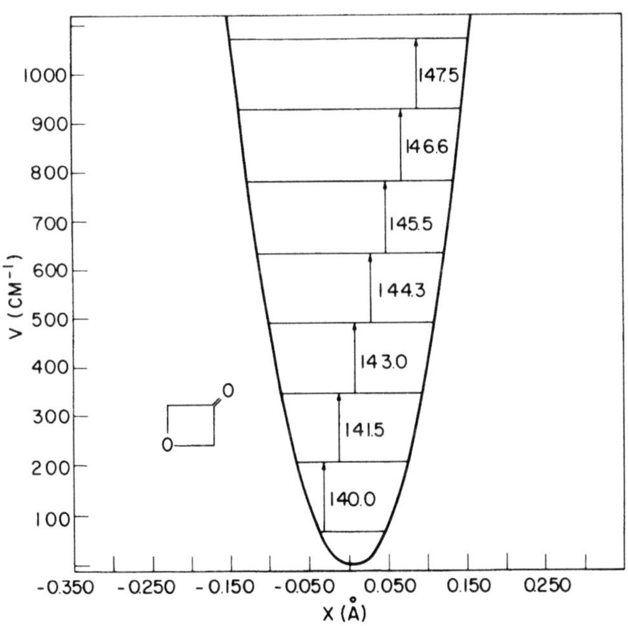

Fig. 4.1. Potential function for the ring-puckering vibration of oxetanone-3. The abscissa is in Å for a coordinate defined as in Fig. 3.1 with the carbonyl moving rigidly with the ring. A reduced mass μ = 151 amu was used.
[Reproduced from Carreira, L. A., Lord, R. C.: J. Chem. Phys. *51*, 3225 (1969).]

The rotational constants, expressed by power series through quartic terms, are averaged over the ring-puckering vibrational states:

$$\beta_v = \beta^\circ + \beta^{(2)} \langle v | Z^2 | v \rangle + \beta^{(4)} \langle v | Z^4 | v \rangle \tag{4.2}$$

where β_v represents the A, B, or C rotational constant in the v'th ring-puckering state. In principle the β°, $\beta^{(2)}$ and $\beta^{(4)}$ are multidimensional Taylor or mixed Fourier/Taylor expansions in the remaining 3N − 7 vibrational modes, averaged over the ground vibrational states of these modes[19]. In practice, the β°, $\beta^{(2)}$ and $\beta^{(4)}$ coefficients are treated as empirical parameters. The rotational constants as a function of vibrational state then depend on these parameters and on the expectation values of the operators Z^4 and Z^2. If Z^2_{ij} and Z^4_{ij} represent these operators in the harmonic oscillator basis, then β_v is given by

$$\beta_v = \beta^\circ + \beta^{(2)} \sum_i \sum_j t_{iv} t_{jv} Z^2_{ij} + \beta^{(4)} \sum_i \sum_j t_{iv} t_{jv} Z^4_{ij} \tag{4.3}$$

The expectation values represented by the double sums in Eq. (4.3) depend on the potential function in Eq. (3.27). For a given harmonic frequency in the basis set, the matrix elements Z^2_{ij} and Z^4_{ij} are fixed but the t_{iv} and t_{jv} depend on the value of B in the dimensionless potential of Eq. (3.27). For a single-minimum potential there is a high degree of correlation between the $\beta^{(4)}$ values and the value of B, each of which leads to curvature in the rotational-constant variation with vibrational state[15]. Since there are ten adjustable parameters, namely, three coefficients for each of the rotational constants plus one potential constant, B, in the reduced potential, it is necessary to determine the rotational constants in a large number of vibrational states if microwave data alone are used.

In the case of oxetanone-3, the coefficients in the rotational constant expansions [Eq. (4.2)] were treated as empirical parameters and the potential function was taken from a previous vibrational study[10]. Figure 2.5 shows the smooth variation, with a definite curvature, of the B rotational constant with ring-puckering vibrational state. Table 4.2 lists the observed and calculated values of the rotational constants. The smooth variation indicates a single-minimum potential with a definite curvature due to the quartic potential term and the quartic terms in the expansion [Eq. (4.3)].

Table 4.2. Comparison of rotational constants in MHz observed and calculated from empirical fit to vibrational potential function for oxetanone-3[15]

	A			B			C		
V	Calc	Obs	Diff	Calc	Obs	Diff	Calc	Obs	Diff
0	12129.34	12129.22	0.12	4956.28	4956.29	−0.01	3686.78	3686.77	0.01
1	12082.76	12082.93	−0.17	4960.37	4960.34	0.03	3696.47	3696.45	0.02
2	12035.87	12035.89	−0.02	4964.83	4964.85	−0.02	3706.08	3706.14	−0.06
3	11988.66	11988.76	−0.10	4969.64	4969.64	0.00	3715.62	3715.61	0.01
4	11941.14	11940.79	0.35	4974.77	4974.77	0.00	3725.09	3725.05	0.04
5	11893.31	11893.39	−0.08	4980.22	4980.22	0.00	3734.49	3734.51	−0.02

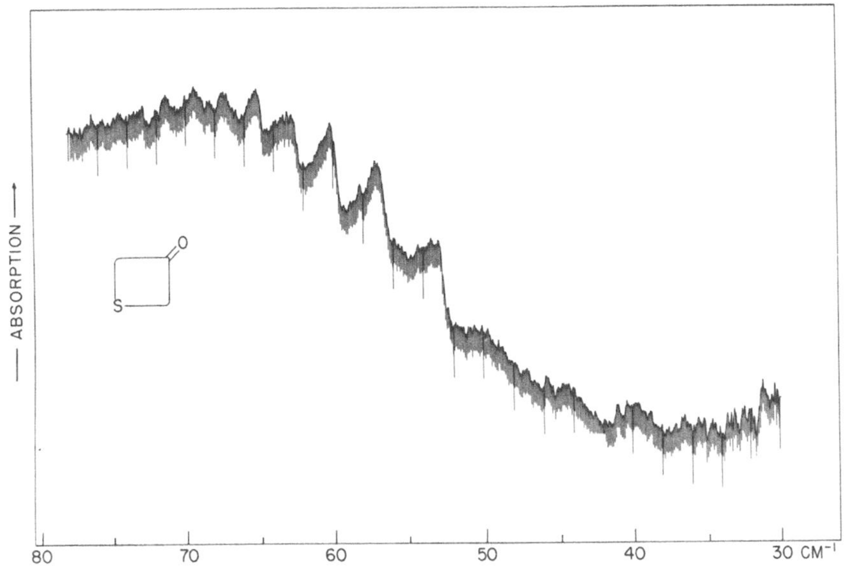

Fig. 4.2. Far-infrared spectrum of thietanone-3. The Q-branch transitions, showing considerable rotational degradation, are observed on the unresolved, overlapped P and R branch transitions. P ~ 1 torr; pathlength = 20 m.
[Reproduced from Blackwell, C. S., Lord, R. C.: J. Mol Spectroscopy *55*, 460 (1975).]

Figure 4.2 shows the far infrared spectrum of the related molecule thietanone-3 reported by Blackwell and Lord[43]. The Q-branches of the *c*-type transitions, superimposed on the overlapped P and R branches, show considerable rotational degradation. The reported frequencies are not the Q-branch maxima but have been approximately corrected for vibration-rotation interaction. The single-minimum potential determined by fitting the far infrared frequencies (Fig. 4.3) shows that the frequencies are considerably lower and the amplitudes of vibration considerably greater than in oxetanone-3. Correspondingly, the quartic anharmonicity is greater.

The microwave spectrum of thietanone-3 was studied by Avirah et al.[44]. In this case, enough microwave data were obtained to determine the dimensionless potential function as well as the expansion coefficients for the rotational constants. Additional data are necessary to determine the energy scale factor, A, in Eq. (3.27). In principle, this may be done by measuring the relative intensities of the microwave lines in different vibrational states, but since the vibrational energy spacings were more accurately obtainable from the far infrared spectra, the latter data were used. The simultaneous fit to 30 rotational constants and the far-infrared transition frequencies is given in Table 4.3. The far-infrared data are reproduced with an rms deviation of 0.24 cm^{-1}. The reduced potential function determined by Blackwell and Lord[43] by fitting the far infrared frequencies alone is

$$V(cm^{-1}) = 9.90(Z^4 + 6.17Z^2) \qquad (4.4a)$$

compared to that determined from both far infrared and microwave data

$$V(cm^{-1}) = 9.81(Z^4 + 6.41Z^2) \qquad (4.4b)$$

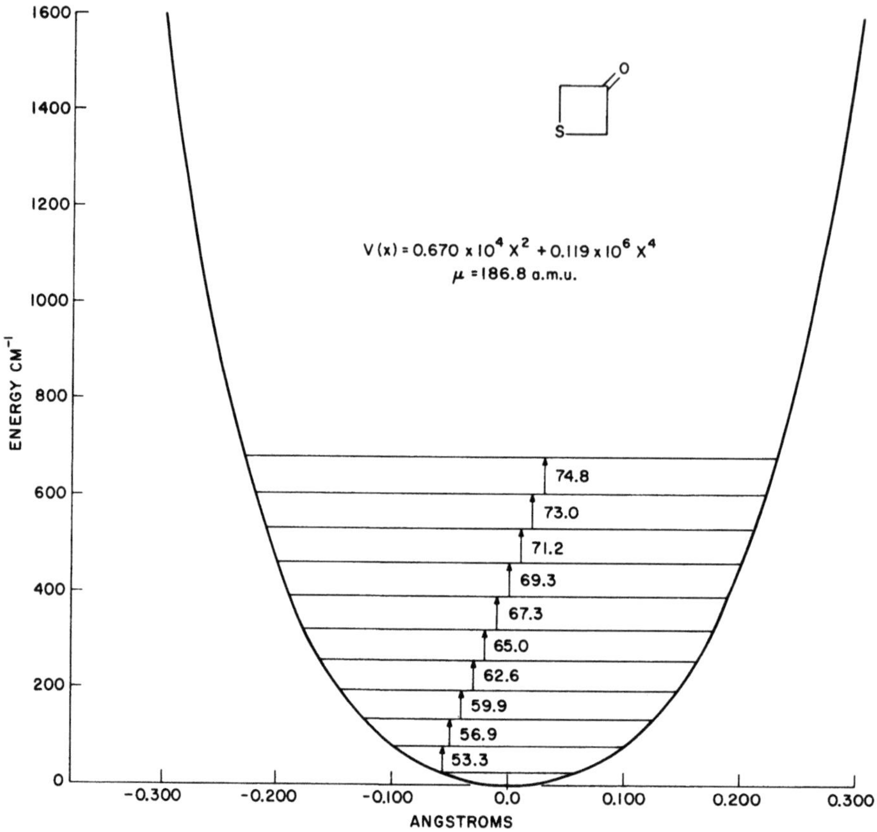

Fig. 4.3. Potential function for the ring-puckering vibration of thietanone-3 determined from fitting the far infrared transitions.
[Reproduced from Blackwell, C. S., Lord, R. C.: J. Mol. Spectroscopy *55*, 460 (1975).]

The two major factors contributing to the ring-puckering potential are angle strain and torsional interactions. In four-membered rings, the valence angles within the ring are generally quite a bit smaller than their values in open chain molecules. Since the ring angles are, on the average, at their maximum possible values for a planar ring, ring strain favors this conformation. On the other hand, torsional interactions generally favor non-planar ring conformation.

For both oxetanone-3 and thietanone-3, angle strain dictates the planar ring conformation. Since there are no adjacent methylene groups, the torsional interactions are not as important for these molecules. On the other hand, the ring-puckering potential for oxetanone yields a higher frequency compared to thietanone than can be accounted for simply on the basis of the difference in the reduced masses. This can be attributed to the fact that there is a considerable difference in the angle strain in the two molecules. A CSC angle in an open chain molecule is generally smaller than the corresponding COC angle. In addition, the force constant for COC bending is considerably greater than for CSC bending. Consequently, the larger amplitude and lower frequency of the ring puckering in thietanone-3 must result from the smaller angle strain than in oxetanone-3.

Table 4.3. Comparison of rotational constants (in MHz) observed and calculated from empirical fit to vibrational potential function $V(Z) = 9.81\ (Z^4 + 6.41\ Z^2)$ thietanone-3[44]

	A		B		C	
V	Obs	Obs−cal	Obs	Obs−cal	Obs	Obs−cal
0	10205.06	−0.33	3266.63	+0.06	2559.70	0.00
1	10117.74	−0.05	3277.54	−0.02	2574.95	+0.04
2	10041.06	+0.58	3287.38	−0.06	2588.56	−0.02
3	9970.36	+0.09	3296.49	−0.08	2601.17	−0.04
4	9905.80	+0.42	3305.13	−0.02	2613.03	−0.03
5	9844.70	+0.01	3313.33	+0.04	2624.32	+0.01
6	9786.86	−0.59	3321.16	+0.07	2635.11	+0.03
7	9732.27	−0.84	3328.67	+0.07	2645.48	+0.05
8	9681.29	+0.03	3335.89	+0.03	2655.47	+0.02
9	9632.24	+0.66	3342.80	−0.10	2665.11	−0.05

$A_v = 10252.29 - 258.12 < Z^2 >_v + 2.89 < Z^4 >_v$
$B_v = 3260.74 + 31.75 < Z^2 >_v + 0.24 < Z^4 >_v$
$C_v = 2551.63 + 43.98 < Z^2 >_v + 0.30 < Z^4 >_v$

(ii) Trimethylene Oxide. That there can be a delicate balance between angle strain and torsional interactions is indicated by the nature of the potential function for trimethylene oxide. This molecule has a double-minimum potential function but with a very small barrier. There are three adjacent methylene groups and torsional interactions play a much greater role than for the two examples given above.

Historically, trimethylene oxide was the first ring molecule for which a ring-puckering potential was determined from spectroscopic data. It has been the most extensively studied ring molecule, having been investigated by far infrared, microwave, Raman and mid infrared techniques[5, 9, 45–59]. Several isotopic species have been synthesized and studied.

During the late 1950's and early 1960's when the initial work on the ring puckering in trimethylene oxide was done, data were much harder to obtain. High-resolution far infrared spectroscopy was in its infancy and Raman spectra of puckering vibrations had not yet been obtained. Today there is a wealth of data available on trimethylene oxide that strikingly demonstrates the success of the simple one-dimensional quartic-quadratic Hamiltonian [Eqs. (3.22), (3.27)]. At the same time, since the data are so extensive, the limitations of the simple one-dimensional potential function can be examined.

The far infrared spectrum of trimethylene oxide is shown in Fig. 2.2. The pattern of transitions is rather regular with the exception of the 0–1 transition which is quite low in both frequency and intensity. The lower intensity is primarily due to the effect of stimulated emission, as shown by the approximate expression for the relative intensities of the various transitions:

$$I_{rel} = \left(e^{-E_v/kT} - e^{-E_{v+1}/kT} \right) \nu_{v \to v+1} <v+1 \,|\, Z \,|\, v>^2 \tag{4.5}$$

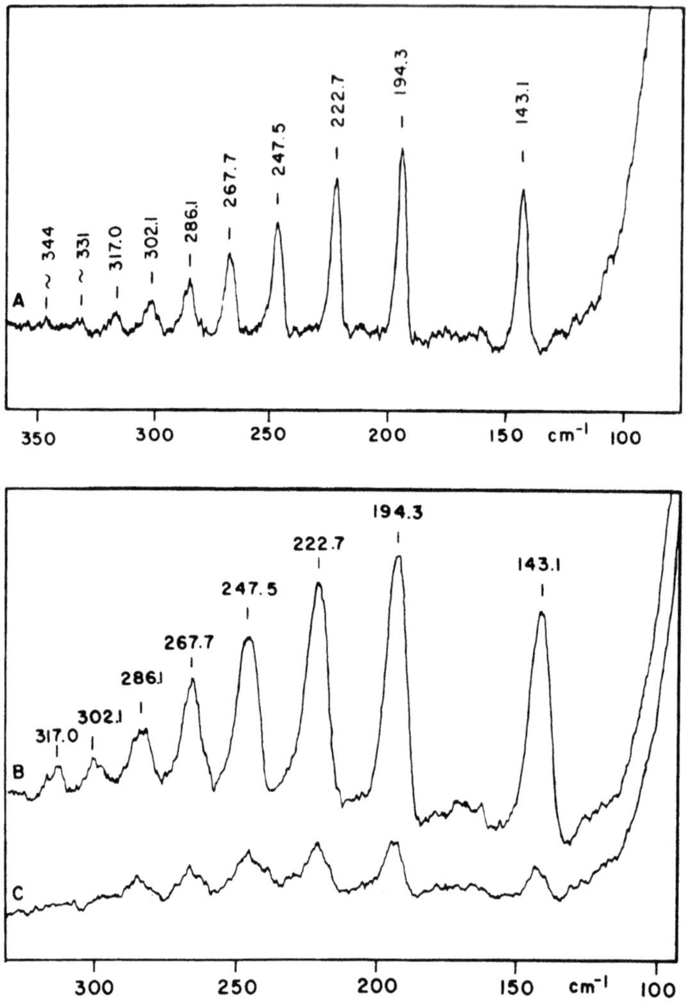

Fig. 4.4. Raman spectrum of trimethylene oxide.
[Reproduced from Kiefer, W., Bernstein, H. J., Danyluk, M., Wieser, H.: Chem. Phys. Letters *12*, 605 (1972).]

Figure 4.4 gives the Raman spectrum of trimethylene oxide[55]. Although the $\Delta v = 1$ transitions are allowed, the prominent features are the $\Delta v = 2$ transitions. Since these overtones are totally symmetric, the sharpness of the Q-branches of such Raman transitions accounts for their prominence in the spectrum.

The irregularity of the position of the 0–1 transition in the far-infrared spectrum or the 0–2 transition in the Raman spectrum compared to the other transitions in the series is indicative of a small barrier in the potential function at the coordinate zero. A small barrier, less than the zero point energy, affects the positions of the levels and the wave functions for the even levels much more than the odd levels[19]. The effect is largest for the zero level and correspondingly less for higher levels.

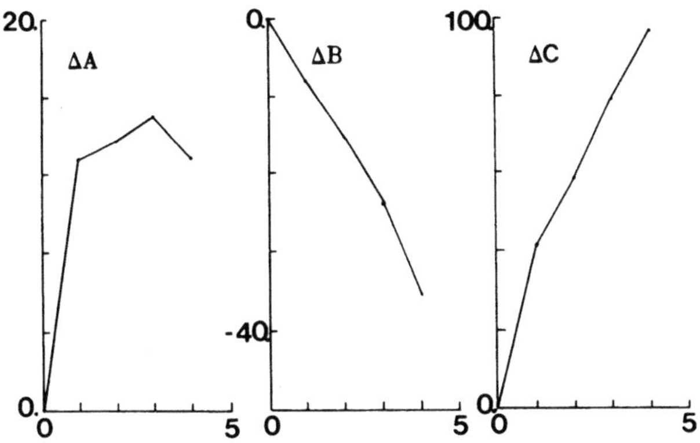

Fig. 4.5. Variation of the rotational constants (in MHz) with ring-puckering vibrational state for trimethylene oxide. The effect of the small barrier is quite dramatic when compared to the smooth variation in Fig. 2.5 for oxetanone-3.
[Reproduced from Chan, S. J., Zinn, J., Fernandez, J., Gwinn, W. D.: J. Chem. Phys. *33*, 1643 (1960).]

Since even functions are more affected than odd functions, the variation of rotational constants with ring-puckering quantum states is expected to deviate from the regular dependence shown by planar molecules (e. g. Fig. 2.5 for oxetanone-3). From Eq. (4.2) we see that alteration of the wave functions of the even levels will affect the expectation values of Z^2 and Z^4 in this equation, primarily Z^2, and lead to an irregular pattern for the lowest few levels.

This was found to be the case for the microwave data for trimethylene oxide[45]. Figure 4.5 indicates the variation of the rotational constants for trimethylene oxide

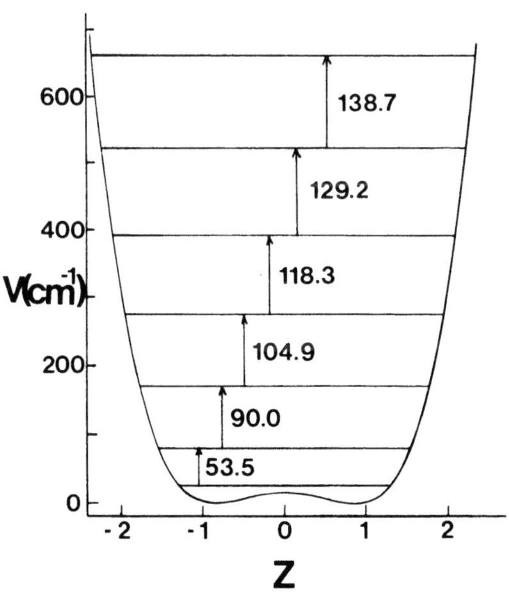

Fig. 4.6. Ring-puckering potential function for trimethylene oxide. The height of the barrier (~15 cm^{-1}) is less than the zero-point energy.

in the ground state and four excited states of the ring-puckering vibration, while in Fig. 4.6 the ring-puckering potential function with a barrier less than the zero point energy is shown. Comparison of the data on oxetanone-3, thietanone-3 and trimethylene oxide indicates that both the pattern of ring-puckering transitions in far infrared or Raman spectra as well as the rotational-constant variation with ring-puckering vibrational state are very sensitive tests of the presence or absence of even very small barriers to planarity.

Extensive studies of the effects of centrifugal distortion in trimethylene oxide and deuterated analogs have been carried out[57, 58]. The distortion constants show a zig-zag dependence on the ring-puckering quantum number similar to that observed for the rotational constants. The results were interpreted according to a simple modification of the standard theory of centrifugal distortion[60–62] in terms of the potential function for the large-amplitude ring-puckering coordinate[57].

A high-resolution far-infrared study (~0.25 cm^{-1}) pointed out another effect on the determination of the ring-puckering potential function[9]. Calculations of the vibration-rotation band contour for a symmetric rotor from the microwave rotational constants allowed determination of the positions of the band origins. These differed from the positions of the Q-branch maxima by 0.1 to 0.25 cm^{-1} to higher frequency. Since the vibrational energy separations were determined more accurately, they were fitted to more significant figures by including a sixth-power potential term in the Hamiltonian

$$H = -\frac{\hbar^2}{2\mu}\frac{d^2}{d\bar{x}^2} + a\bar{x}^4 + b\bar{x}^2 + c\bar{x}^6 \qquad (4.6)$$

The barrier[9] was reported as 15.52 ± 0.05 cm^{-1} compared to 15.3 ± 0.5[47] and 15.1 ± 0.5 cm^{-1} [52] reported previously.

The stated barrier uncertainty of ±0.05 cm^{-1} is somewhat misleading in that effects which can account for several cm^{-1} in the barrier have been neglected. Use of the coordinate $\bar{x}$ with the assumption of a constant effective mass can change the barrier by 2–5 cm^{-1} and certainly has at least as large an effect as the inclusion of a sixth power term in the potential. Another thing that should be kept in mind is that the ring-puckering potential is an effective potential containing contributions from averaging anharmonic interaction terms over the zero-point vibrations of the other $3N-7$ vibrational modes [see Eqs. (3.18) to (3.20)]. In the case of ring molecules, these contributions may be of the order of a few cm^{-1}. This latter effect is analogous to the zero-point vibrational contribution to the determination of effective rotational constants. With the above in mind, it is seen that there is dubious physical significance to the value of the coefficient of the sixth power term in the potential function [Eq. (4.6)] determined from fitting the data.

The effect of the zero-point averaging over the other modes manifests itself in two ways. A number of progressions of combination and difference bands have been observed in the mid-infrared and Raman spectra of trimethylene oxide. These have been studied extensively by Wieser and co-workers[50–56], who found small changes in the ring-puckering intervals on excitation of quanta of higher frequency vibrational modes. They also studied the spectra of a number of deuterated analogues of trimethylene oxide and found an isotopic dependence of the ring-puckering potential functions.

This dependence may be interpreted as due to several different effects. If the ratio of the reduced masses of the isotopic species remains constant or nearly so as the vibrational amplitude changes, neglect of the dependence of the reduced mass on coordinate will not introduce differences between the ring-puckering potential functions determined for isotopic species [Eq. (3.21)]. However, cross terms in the kinetic energy and harmonic cross terms in the potential energy between the ring-puckering vibration and higher frequency modes of the same symmetry can lead to different potential functions for isotopic species. If we consider removing the harmonic cross terms in the kinetic and potential energy by a "normal coordinate transformation", the form of the lowest-frequency coordinate may differ for isotopic species [Eqs. (3.14) to (3.17)]. Consequently, we would not expect the same potential function for the different species.

Probably of more importance are the contributions of anharmonic interaction terms [Eq. (3.19)]. Since the zero-point contribution of these terms will be different for the various isotopic species, a difference in the effective potential is expected. If we consider only the effect of the anharmonic interaction terms up to fourth degree through first order, the following result is obtained

$$V_{eff}(\bar{x}) \approx a\bar{x}^4 + \left(b + \sum_{i=1}^{3N-7} a_{ix} \langle v_i | q_i^2 | v_i \rangle \right) \bar{x}^2 \tag{4.7}$$

where a and b are (approximately) invariant to isotopic substitution and the expectation values of the squares of the 3N–7 high frequency coordinates are not. That these types of interaction terms are probably the most important may be seen from examining Table 4.4 from the work of Wieser and co-workers[50–52]. It is seen that the quartic terms in the dimensioned potential are essentially constant for the isotopic species while the variation in the quadratic coefficient b is more pronounced. This is the result expected from the considerations leading to Eq. (4.7).

In principle, it should be possible to obtain enough data to correct the effective potential function for trimethylene oxide to a "vibrationless" state. This potential function should then be isotopically invariant. This may require determination of the ring-puckering intervals in the excited states of the other 3N–7 modes and

Table 4.4. Comparison of the potential functions for isotopic species of trimethylene oxide[52)]

	TMO-d_0	α-d_2	β-d_2	α,α'-d_4	d_6
A(cm^{-1})[a]	28.12	25.46	26.10	22.85	21.54
B	−1.465	−1.445	−1.465	−1.445	−1.445
a(10^5 cm^{-1} Å^{-4})[b]	7.16	7.07	7.19	7.07	7.07
b(10^3 cm^{-1} Å^{-2})	−6.58	−6.13	−6.34	−5.81	−5.64
μ (amu)	95.7	110.4	107.3	129.8	141.8

[a] Equation (3.27). [b] Equation (3.22).

treatment of perturbations of particular levels (e. g. Fermi resonance) to remove these effects from the determination of the effective potentials. Despite the voluminous data on trimethylene oxide, this point has not yet been reached.

The failure to remove all of the vibrational averaging effects is not too upsetting. The ability to determine a potential function with a barrier accurate to within a few cm^{-1} is rather remarkable. The fact that the data can reveal even a very small barrier – in the case of trimethylene oxide, smaller than the zero point energy – shows that they are a very sensitive probe of the molecular dynamics.

(iii) Trimethylene Sulfide and Methylenecyclobutane. Shortly after the original work on trimethylene oxide, the results of a microwave study on trimethylene sulfide were published[18]. In this case, the potential function has a much higher barrier, 274 cm^{-1}, and the effects on the spectra are quite dramatic. The potential function, with some of the calculated vibrational spacings[18], is shown in Fig. 4.7. In contrast to the regular pattern of transitions converging to higher frequency observed for planar molecules, the frequency pattern is more complicated, until levels well above the barrier are reached. Transitions with $\Delta v = 1$ and $\Delta v = 3$ are observed in the far-infrared spectrum[63, 64], while the prominent transitions in the Raman spectrum are the Q-branches for the totally symmetric $\Delta v = 2$ transitions[65].

The effect of the ring-puckering vibration, particularly in the $v = 0$ and $v = 1$ states, on the rotational spectrum of trimethylene sulfide is striking[18]. In Eq. (3.10) the terms involving the ring-puckering momentum operator and components of the rotational angular momentum operators are no longer small when vibrational energy spacings become of the same order as the rotational energy spacings of interest. Thus non-rigid rotor spectra result and a more complete Hamil-

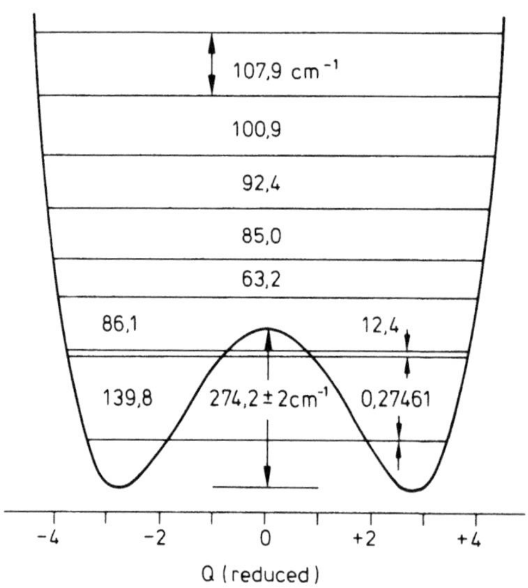

Fig. 4.7. Ring-puckering potential function for trimethylene sulfide. The barrier at the planar conformation is 274 cm^{-1}.
[Reproduced from Harris, D. O., Harrington, H. W., Luntz, A. C., Gwinn, W. D.: J. Chem. Phys. *44*, 3467 (1966).]

tonian including vibration-rotation cross terms is required. There are two physically equivalent types of vibration-rotation interaction terms which arise. Their relative magnitudes depend on the choice of the rotating axis system used to set up the Hamiltonian[18, 20].

If we restrict the Hamiltonian given by Eq. (3.10) to one vibrational coordinate, a ring-puckering coordinate with a constant effective mass, the following vibration-rotation Hamiltonian results

$$H = \frac{1}{2}\Bigg\{ g_{aa}(\overline{x})P_a^2 + g_{bb}(\overline{x})P_b^2 + g_{cc}(\overline{x})P_c^2 + g_{ac}(\overline{x})(P_aP_c + P_cP_a) + \left[2g_{bx}(\overline{x})p_x + \frac{\hbar}{i}\left(\frac{\partial g_{bx}}{\partial x}\right)\right] P_b + g_{xx}P_x^2 \Bigg\} + a\overline{x}^4 + b\overline{x}^2 \tag{4.8}$$

The two types of vibration-rotation interaction terms are the $P_aP_c + P_cP_a$ term and the P_b term (the *b* axis is perpendicular to the symmetry plane which is maintained throughout the ring-puckering.). Their coefficients are functions of the vibrational coordinate or the vibrational momentum or both. It is possible to choose the coordinate system so that the $P_aP_c + P_cP_a$ term is zero and all of the coupling between rotational angular momentum and vibrational momentum is manifested by the P_b term. The matrix elements for the Hamiltonian in the basis of the solutions to the pure vibrational (J = 0) problem, are

$$H_{vv} = E_v + A_vP_a^2 + B_vP_b^2 + C_vP_c^2 \tag{4.9a}$$

and

$$H_{vv'} = F_{vv'}P_b \tag{4.9b}$$

E_v is the appropriate vibrational eigenvalue. A_v, B_v and C_v are given by

$$\beta_v = \frac{1}{2}\langle v|\, g_{\beta\beta}(\overline{x})\,|v\rangle \tag{4.10a}$$

β = A, B, or C

$$F_{vv'} = \langle v|\, g_{bx}(\overline{x})p_x\,|v'\rangle \tag{4.10b}$$

where the contribution of $\frac{\hbar}{i}\left(\frac{\partial g_{bx}}{\partial x}\right)$ has been neglected.

The off-diagonal coupling terms, Eqs. (4.9b) and (4.10b), may be treated as perturbations making a small contribution to the effective rotational constants in the various vibrational states for all cases except those where the vibrational energy spacing is comparable to typical rotational energy spacings. In the case of trimethylene sulfide, the 0–1 vibrational spacing is comparable to low-J rotational spacings and only for this pair of levels is the Hamiltonian described by Eqs. (4.9a, b) treated explicitly. The parameters determined from the microwave data include the three effective rotational constants for each of the two vibrational states, the 0–1 vibrational interval and the vibration-rotation interaction constant, F_{01}[18]. The rotational spectra in the higher ring-puckering states were rigid rotor spectra and were so treated. Figure 2.7 indicates vibration-rotation levels for the v = 0 and v = 1 states which are affected by the cross term, Eq. (4.9b).

As was pointed out by Butcher and Costain in their work on cyclopentene[66], the effective rotational constants for the 0 and 1 states contain contributions from the cross term Eq. (4.9b) different from those to the effective rotational constants in the higher states. Using second order perturbation theory, Scharpen derived these corrections to the rotational constants[20]. Pickett[21] considered a different choice of rotation axes, for which all of the vibration-rotation interaction was expressed in the coordinate dependence of the off-diagonal term g_{ac} [Eq. (4.8)]. The resulting diagonal Hamiltonian matrix element is the same as in Eq. (4.9a), while the off-diagonal term is given by

$$H_{vv'} = F'_{vv'}(P_aP_c + P_cP_a) \qquad (4.11a)$$

where

$$F'_{vv'} = \frac{1}{2} <v| g_{ac}(\bar{x}) |v'> \qquad (4.11b)$$

With this Hamiltonian, Pickett[21] derived the same 0–1 vibrational splitting as that found earlier by Harris et al.[18] and the same effective rotational constants as those obtained by Scharpen using perturbation theory[20].

The simple constant-effective-mass, quartic-quadratic Hamiltonian, Eqs. (3.22), (3.27), was found quite adequate to reproduce the observed far infrared transitions, account for the rotational constant variation [via Eq. (4.2)] and faithfully reproduce the 0–1 inversion splitting derived from the vibration-rotation interaction analysis. As with trimethylene oxide, Wieser et al. have studied a number of deuterated derivatives of trimethylene sulfide[67–69]. The barriers derived vary over a range of $\sim 8\ cm^{-1}$. This variation is probably due to the factors mentioned above for trimethylene oxide and gives an indication of the precision to which barriers may be determined using a simple effective one-dimensional Hamiltonian.

Another molecule with a similar double-minimum potential function is methylenecyclobutane. Again, the 0–1 inversion splitting is of the order of typical rotational energy spacings and non-rigid rotor spectra result[17]. The Hamiltonian given by Eqs. (4.9a, b) was used to fit the data for the v = 0 and v = 1 states. After suitable correction of the rotational constants for these states, they were used along with the rigid-rotor constants for v = 2–6 to determine the potential function. The 0–1 splitting was used to scale the function and a barrier of $160 \pm 40\ cm^{-1}$ was reported. Figure 2.6 shows the variation of the A rotational constant for methylenecyclobutane and for comparison, cyclobutanone[16]. The zig-zag pattern is quite evident for methy-

Table 4.5. Assigned rotational transition frequencies[a] for the v = 0 and v = 1 ring-puckering states of methylenecyclobutane[17]

Transition	v = 0 v(Obs)	Pert.[b]	v = 1 v(Obs)	Pert.[b]
$0_{00} \rightarrow 1_{01}$	8066.56(−0.01)	−11.37	8072.57(−0.15)	+17.26
$1_{01} \rightarrow 2_{02}$	15979.45(−0.04)	−18.61	15992.70(−0.16)	+32.17
$1_{11} \rightarrow 2_{12}$	14989.22(−0.05)	−8.48	15017.02(−0.19)	+35.78
$1_{10} \rightarrow 2_{11}$	17270.27(+0.04)	−43.70	17295.12(−0.35)	+55.19
$2_{02} \rightarrow 3_{03}$	23602.37(−0.17)	−18.79	23625.70(−0.05)	+43.19
$2_{20} \rightarrow 3_{21}$	24772.49(+0.24)	−73.81	25053.92(−0.76)	+304.68
$2_{21} \rightarrow 3_{22}$	24160.30(+0.32)	−73.14	24541.66(−0.76)	+375.43
$2_{12} \rightarrow 3_{12}$	22393.03(+0.02)	−10.73	22444.99(−0.19)	+61.44
$2_{11} \rightarrow 3_{12}$	25799.33(−0.04)	−65.42	25839.36(−0.26)	+80.54
$3_{03} \rightarrow 4_{04}$	30870.65(−0.27)	−12.78	30910.21(+0.37)	+53.28
$3_{21} \rightarrow 4_{22}$	33506.17(−0.03)	−109.77	33714.43(+0.63)	+249.10
$3_{22} \rightarrow 4_{23}$	32064.71(+0.37)	−123.29	33599.42(+0.73)	+496.66
$3_{13} \rightarrow 4_{14}$	29705.05(+0.07)	−11.27	29803.94(−0.05)	+107.61
$3_{31} \rightarrow 4_{32}$	32092.58(−0.50)	−498.54	32407.05(+0.07)	−79.33
$3_{12} \rightarrow 4_{12}$	34179.86(−0.21)	−88.88	34243.14(+0.33)	+104.71
$3_{30} \rightarrow 4_{31}$	32445.01(+0.48)	−226.69	32482.60(+0.38)	−78.04
$4_{04} \rightarrow 5_{05}$	37846,63(−0.14)	−5.58	37920.11(+0.93)	+77.44
$4_{22} \rightarrow 5_{23}$	42420.05(−0.77)	−150.00	42564.68(+2.40)	+198.74
$4_{40} \rightarrow 5_{41}$	40858.29(+1.12)	+111.96	40577.72(+0.37)	−37.32
$4_{13} \rightarrow 5_{14}$	42324.34(−0.44)	−122.63	42433.65(+1.48)	+129.23
$4_{31} \rightarrow 5_{32}$	40909.99(−1.77)	−170.74	40827.63(−0.26)	−99.73
$4_{23} \rightarrow 5_{24}$	39792.83(+0.39)	−246.23	40654.28(+2.71)	+713.90
$4_{41} \rightarrow 5_{42}$	40844.02(+0.92)	+105.18	40571.01(−0.09)	−37.56
$4_{14} \rightarrow 5_{15}$	36919.56(+0.15)	−10.41	37141.00(+0.26)	+227.21
$4_{32} \rightarrow 5_{33}$	40086.12(−2.08)	−720.43	40572.65(−0.23)	−100.28

a Frequencies are in megahertz with an estimated uncertainty of ± 0.05 MHz[17]. The quantity in parentheses after each frequency is the calculated frequency minus the observed frequency.
b Calculated frequency minus the calculated rigid-rotor frequency.

lenecyclobutane. Cyclobutanone has a small barrier, ~ 1/2 the barrier in trimethylene oxide, and shows a distinct irregularity in the dependence of the rotational constants on vibrational state. This figure illustrates the extreme sensitivity of the microwave data to a very small barrier to planarity. Similarly, the irregularity of the frequency pattern in the far-infrared and Raman spectra provides a test of the planarity or non-planarity of the ring. Table 4.5 lists the observed and calculated microwave frequencies for the v = 0 and v = 1 states of methylenecyclobutane. The vibration-rotation contribution to the frequencies is seen to be substantial.

Figure 4.8 shows the ring-puckering transitions with v = 2 observed in the Raman spectrum of methylenecyclobutane[65]. The fit to the Raman data yields a potential function with a barrier of 140 ± 5 cm^{-1}. The Raman lines and their assignments were used to account for combination bands involving the ring-puckering vibration observed in the mid infrared spectrum[70].

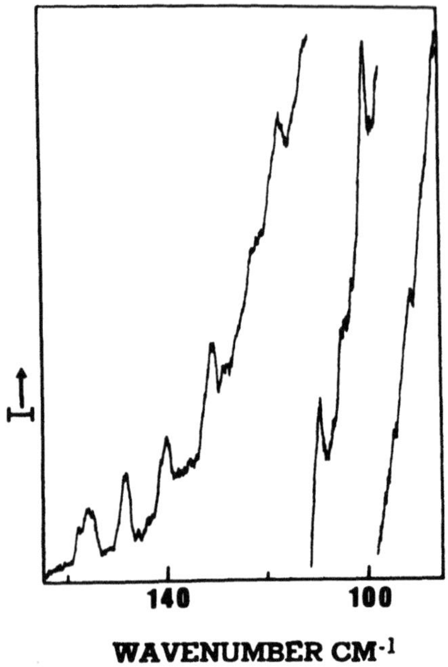

Fig. 4.8. Raman spectrum of methylenecyclobutane. The observed maxima are Q branch transitions of the totally symmetric $\Delta v = 2$ ring puckering transition [Reproduced from Durig, J. R., Shing, A. C., Carreira, L. A., Li, Y. S.: J. Chem. Phys. *57*, 4398 (1972).]

(iv) Silacyclobutane. Silacyclobutane exhibits an extensive ring-puckering spectrum in the far-infrared (Fig. 4.9). This spectrum, reported by Laane and Lord[71], has transitions assigned as $\Delta v = 1$, $\Delta v = 2$, $\Delta v = 3$ and one tentative assignment of a $\Delta v = 4$ transition. These data were fitted with the two-parameter quartic-quadratic Hamiltonian given by Eq. (3.27). The observed and calculated transition frequencies along with their intensities are given in Table 4.6. That such extensive data are fitted so well with the two-parameter Hamiltonian is a remarkable success for the simple

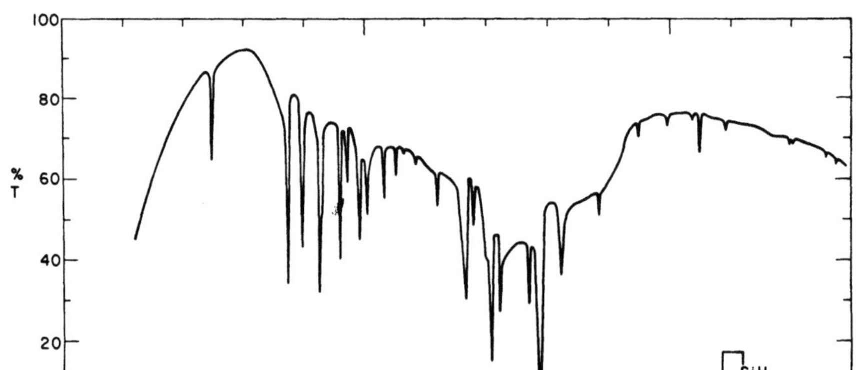

Fig. 4.9. Far infrared spectrum of silacyclobutane. P = 60 torr, pathlength, 8 m [Reproduced from Laane, J., Lord, R. C.: J. Chem. Phys. *48*, 1508 (1968).]

Table 4.6. Observed and calculated far-infrared transitions for silacyclobutane[71]

Transition	Frequency (cm^{-1})		Δ	Relative Absorbance[a]	
	Calculated	Observed		Calculated	Observed
0– 1		(0.003)[b]	–	10^{-9}	–
1– 2	159.48			2.50	
0– 3	159.74			2.50	2.3
		157.78	1.70		
0– 2	159.48			0.32	
1– 3	159.75			0.32	
2– 3	0.264	(0.26)[b]	0.00	10^{-4}	–
3– 4	133.38			1.31	0.5
		133.45	–0.07		
2– 4	133.64			.17	
2– 5	142.26			1.96	0.8
		141.80	0.46		
3– 5	142.00			0.19	
4– 5	8.63	–	–	0.05	
5– 6	84.90	85.37	–0.47	1.15	0.8
6– 7	50.46	49.85	0.61	0.57	0.3
7– 8	74.67	74.70	–0.03	(1.0)	(1.0)
8– 9	78.99	79.22	–0.23	0.88	0.6
9–10	86.29	86.20	0.09	0.76	~0.3
10–11	91.97	92.10	–0.13	0.59	0.6
11–12	97.22	98.50	–1.28	0.44	0.3
12–13	102.02	101.02	1.00	0.32	0.2
13–14	106.47	106.36	0.11	0.22	0.2
14–15	110.78	110.46	0.32	0.14	0.1
15–16	114.69	113.23	1.46	0.10	0.01
16–17	118.50	117.08	1.42	0.06	0.02
4– 6	93.53	94.41	–0.88	0.06	0.2
5– 7	135.36	135.79	–0.43	0.14	0.2
6– 8	125.14	124.17	0.97	0.08	0.1
7– 9	153.66	153.81	–0.15	0.10	0.3
8–10	165.28	164.68	0.60	0.09	0.3
9–11	178.26	177.11	1.15	0.08	0.15
10–12	189.19	190.21	–1.02	0.06	0.09
11–13	199.24	199.21	0.03	0.05	0.07
12–14	208.49	207.24	1.25	0.03	0.05
1– 4	293.12	291.7	1.4	0.12	vw
3– 6	226.90	~227	–	0.36	~.05
4– 7	144.00	144.56	–0.56	0.96	0.4
5– 8	210.04	210.68	–0.64	0.48	0.20
6– 9	204.12	–	–	–	–
7–10	239.94	239.64		–	0.06
		241.64			0.04
8–11	257.24	256.7	0.5	–	0.01
4– 8?	220.66	218.83	1.83	–	0.11

[a] Relative absorbances calculated with $\partial\mu_a/\partial Q = 0.1$, $\partial^2\mu_c/\partial Q^2 = 0.2$.
[b] Approximate value from microwave work[74].

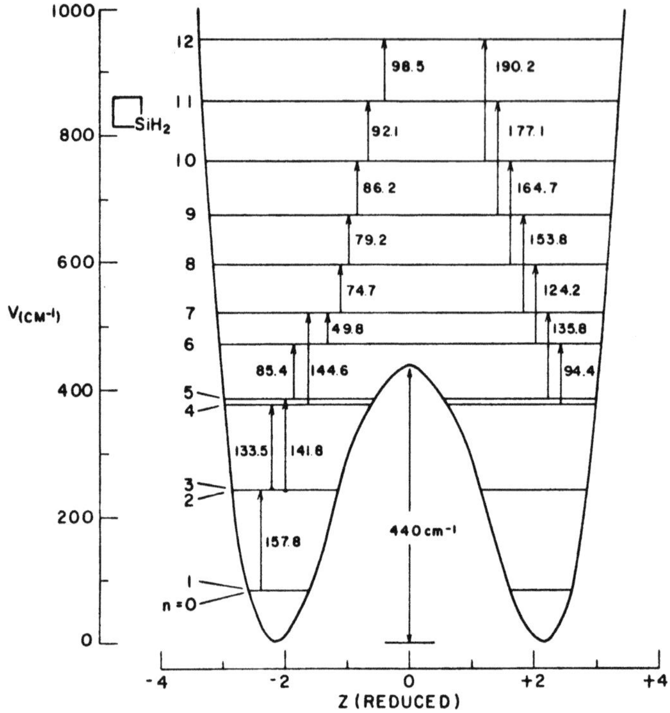

Fig. 4.10. Ring-puckering potential function for silacyclobutane. [Reproduced from Laane, J., Lord, R. C.: J. Chem. Phys. *48*, 1508 (1968).]

one-dimensional model. The potential function derived from the data is shown in Fig. 4.10, yielding a barrier to planarity of 440 cm^{-1}.

The Raman spectrum of gaseous silacyclobutane[72)] confirms the far infrared results. The prominent transitions are those for which $\Delta v = 2$. Combination and difference band progressions involving the ring-puckering vibration and a SiH_2 stretching mode were observed in the mid infrared spectrum[73)]. A small change was found in the effective ring-puckering potential in the first excited state of this mode.

The microwave spectrum of silacyclobutane was studied by Pringle[74)]. In this case, the 0–1 vibrational spacing is smaller (ca. 75 MHz) than typical rotational energy spacings and rigid-rotor spectra with identical rotational constants result for these two states. However, due to the near degeneracy of the vibrational states, the Stark effect of certain lines, particularly those involving J_{0J} or J_{1J} rotational sublevels, is affected drastically. Analysis of the Stark-effect data yielded the 0–1 inversion splitting and the transition moment matrix element $|\langle 0 | \mu_c | 1 \rangle|$. Rovibrational transitions with $\Delta v = 1$ were then observed within 1 MHz of the predicted position. This led to an accurate determination of the 0–1 splitting, 75.75 ± 0.03 MHz.

The 2–3 splitting is of the order of a typical low-J rotational spacing and nonrigid-rotor spectra result. The rovibrational levels were therefore computed from the Hamiltonian of Eqs. (4.9a, b) with the help of a second-order perturbation correction used by Butcher and Costain[66)] for cyclopentene rather than by direct matrix diagonalization. This allowed determination of the 2–3 vibrational spacing as 7793 ± 7 MHz. In principle, the determination of the 0–1 and 2–3 vibrational intervals is sufficient to evaluate the two parameters in the potential function given in Eq. (3.27) and thus also the barrier to planarity. In fact, if this is done, the potential function computed has a barrier of 229 cm^{-1}, in error by almost 50%!

On the other hand, the far infrared data determine a potential function which predicts the 0–1 and 2–3 splittings to within 4% and 3% respectively. Table 4.7 lists a series of calculations due to Pringle. He reports the frequencies calculated by fitting far infrared data alone and the microwave data alone. He also has fitted both microwave and far infrared data together, using a four-parameter potential function successfully employed earlier to fit simultaneously the far-infrared and microwave data for cyclobutanone[74],

$$H = -\frac{\hbar^2}{2\mu}\frac{d^2}{d\bar{x}^2} + a\bar{x}^4 + b\bar{x}^2 + ce^{-d\bar{x}^2} \tag{4.12}$$

In contrast to the case of cyclobutanone, the addition of two more adjustable parameters does not seem warranted in the case of silacyclobutane in that only a small improvement in the fit results. The barrier determined is 442 cm^{-1}, within 2 cm^{-1} of the barrier determined from the simpler quartic-quadratic potential function. As pointed out by Pringle, the tendency is to weight the microwave data heavily because of the precision of the rotational data compared to that of the measurement of the vibrational intervals in the far-infrared or Raman spectrum. However, in doing so, one fails to recognize the limitations of the Hamiltonian. If the potential func-

Table 4.7. Ring-puckering frequencies for silacyclobutane[74]

		Calc frequency $A(Z^4 + BZ^2)$		$A(Z^4 + BZ^2 + Ce^{-DZ^2})$
Vibrational transition	Obs frequency	FIR data[a]	Microwave data[b]	Combined data[c]
1 ← 0	75.75 MHz	72.0 MHz	73.75 MHz	75.68 MHz
3 ← 2	7793.0 MHz	8033.0 MHz	7800.0 MHz	7955.0 MHz
2 ← 1	157.8 cm^{-1}	157.9	86.7	157.3
5 ← 2	141.8	142.0	77.6	141.5
4 ← 3	133.5	133.6	70.0	132.8
6 ← 5	85.4	85.0	44.0	85.4
7 ← 6	49.9	50.7	31.7	49.4
8 ← 7	74.7	74.9	***	74.3
9 ← 8	79.3	79.3	***	78.2
10 ← 9	86.0	86.6	***	85.6
14 ← 13	106.3	106.9	***	107.9

[a] Least squares fit to ir data. Barrier height is 440 cm^{-1}.
[b] Least squares fit to microwave data alone. Barrier height is 229 cm^{-1}.
[c] Gaussian term added to potential, and both sets of data used in least squares fit. Barrier height is 442 cm^{-1}.

tion obtained by fitting very precise rotational data over a limited energy range to an approximate Hamiltonian is inconsistent with a large amount of less precise vibrational data extending over a wide energy range, one may conclude that the Hamiltonian has been pushed beyond its limitations in such a fitting.

It had been suggested that determination of an inversion splitting for a double-minimum potential function, along with the variation of rotational constants, was sufficient to determine the barrier accurately. This looked particularly promising in light of the success of this procedure in the case of trimethylene sulfide[18]. However, subsequent studies showed mixed results. In the microwave study of trimethylene selenide[75], the 2–3 inversion splitting was evaluated by analyzing the vibration-rotation interaction for these two states. The rotational constant variation was used to compute a dimensionless potential function. The far-infrared transition frequencies published earlier by Harvey et al.[76] were then employed to evaluate the scale factor for this reduced potential function. Scaled in such a fashion, the barrier for the potential function calculated from the rotational constant variation was 383 ± 4 cm^{-1}, compared to a barrier of 378 ± 4 cm^{-1} determined previously by Harvey et al.[76] by fitting the far-infrared data alone. *If* the derived 2–3 inversion splitting had been used to scale the microwave potential function, a barrier of 297 cm^{-1}, in error by more than 20%, would have been derived. Perhaps this would be improved by using the four-parameter potential function with the Gaussian barrier [Eq. (4.12)]. However, a similar procedure was not of much help in the case of silacyclobutane.

At any rate, this failure of the simple two-parameter quartic-quadratic Hamiltonian with a constant reduced mass to reproduce simultaneously and precisely the inversion splittings and far-infrared or Raman data should not be considered a serious drawback. Attempts to use this as an indication of a real difference in the shape of the potential function fail to take into account other effects which have been neglected, among them the dependence of the reduced mass on the coordinate.

(v) Cyclobutane. Cyclobutane is the parent hydrocarbon for saturated four-membered ring molecules. Due to the fact that it has no permanent dipole moment, no microwave study is possible. Direct observation of the ring-puckering transitions in the far infrared is also precluded by the fact that this mode is infrared inactive. The first estimate of the barrier to planarity from spectroscopic data was made from the observation of a progression of ring-puckering difference bands from a CH_2-stretching fundamental by Ueda and Shimanouchi[77]. Stone and Mills[33] and, independently, Miller and Capwell[13] observed both combination and difference band progressions of a B_2 CH_2 scissoring mode in the mid infrared spectrum. From these data, by means of combination differences, it was possible to derive the ring-puckering vibrational spacings in both the $v = 0$ and $v = 1$ states of the CH_2-scissoring mode. Similar data were obtained for cyclobutane-d_8. In addition, $\Delta v = 2$ transitions were directly observed in the Raman spectra of both light and heavy cyclobutane[13].

The low-frequency Raman spectra of gaseous cyclobutane and cyclobutane-d_8 are shown in Fig. 4.11, and the mid infrared spectrum in the region of the CH_2 scissoring mode, with the ring-puckering fine structure, is given in Fig. 2.3. Due to minor discrepancies between the assignments of Stone and Mills and those of Miller and Capwell, as well as slightly different methods of deriving and fitting the data,

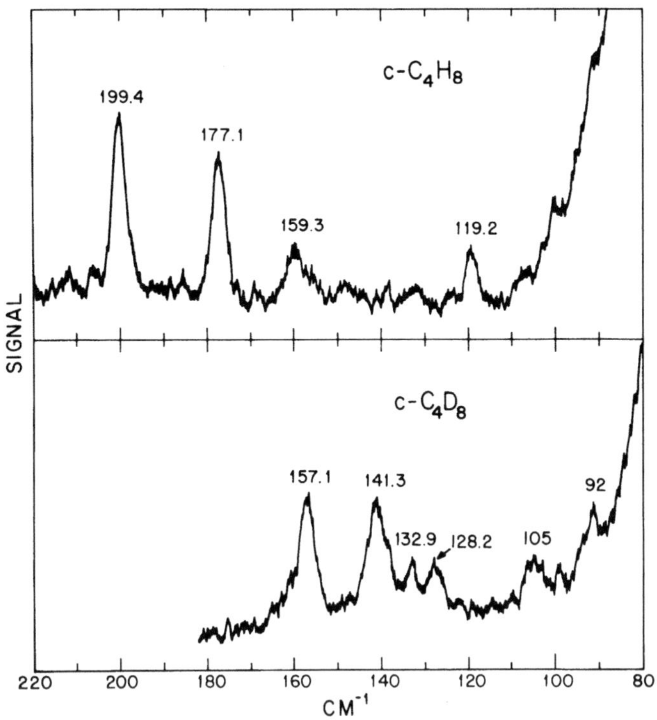

Fig. 4.11. Raman spectra of gaseous cyclobutane and cyclobutane-d_8. The observed Q-branch transitions are the $\Delta v = 2$ ring-puckering transitions.
[Reproduced from Miller, F. A., Capwell, R. J.: Spectrochim. Acta *27A*, 947 (1971).]

slightly different barriers were determined. Stone and Mills derived the same potential function for cyclobutane and cyclobutane-d_8 by adjusting the reduced-mass ratio, but allowed the effective potential functions to depend on the quantum state of the CH_2 (CD_2) scissoring mode. They found that the reduced-mass ratio calculated for a coordinate for which the CH_2 (CD_2) groups move rigidly with the ring considerably underestimated the isotopic shift for the ring puckering. They fixed the reduced mass for cyclobutane at the value for this semirigid model, and empirically adjusted the value for cyclobutane-d_8. On the other hand, Miller and Capwell allowed for an isotopic dependence of the effective potential but not for the dependence of the ring-puckering eigenvalues on the quantum state of the scissoring mode.

Later, Malloy and Lafferty[27)] made minor reassignments of one of Stone and Mills's and two of Miller and Capwell's bands to reconcile the two sets of data. They fitted the infrared and Raman data for cyclobutane, allowing different effective potential functions in the ground and excited states and using only lines which were free from overlap with other lines. The resulting calculated mid infrared puckering structure is compared to the observed in Table 4.8. The data for cyclobutane-d_8 were fitted in the same way with the results given in Table 4.9. Table 4.10 summarizes the potential constants derived [Eq. (3.27)], the barrier heights and the band origins for the CH_2 (CD_2) modes. It is seen that the effective barriers for the ground states

Table 4.8. Puckering structure on ν_{14} band of C_4H_8[27]

Transition	$\nu_{obs}{}^a(cm^{-1})$	$\nu_{calc}(cm^{-1})$	$\nu_{obs}-\nu_{calc}(cm^{-1})$
12–13	1318.9	1319.59	−0.69
11–12	1324.7	1325.23	−0.53
10–11	1331.8	1331.34	+0.46
9–10	1338.2	1337.98	+0.22
8– 9	1346.1	1345.77	+0.33
7– 8	1352.5^b	1352.61^b	−0.11^b
5– 6	1352.2^b	1351.63^b	+1.57^b
6– 7	1373.2	1372.48	+0.72
4– 5	1430.0	1429.55	+0.45
2– 3	1451.3	1451.76	−0.46
1– 2		1455.70	–
0– 1	1453.3^b	1454.60	–
1– 0		1454.62	–
5– 4	1470.5	1470.22	+0.28
7– 6	1527.7	1527.93	−0.23
6– 5	1547.8	1548.51	+0.29
8– 7	1550.3	1550.50	−0.20
9– 8	1558.5	1558.53	+0.03
10– 9	1567.8	1567.56	+0.24
11–10	1574.9	1575.21	−0.31
12–11	1582.4	1582.24	+0.16
13–12	1589.2	1588.72	+0.48
5– 2	1605.8	1606.71	−0.91
3– 4	1299.9^c	1294.70^c	–
7– 4		1646.527	–
2– 1	1649.9^b	1651.432	–
3– 0		1652.351	–

[a] Data taken from Ref.[33]. [b] Blended or overlapped lines not used in fit.
[c] Suspect assignment; not included in fit.

of cyclobutane and cyclobutane-d_8 differ by ~15 cm^{-1}. As shown by Malloy and Lafferty[27], this difference cannot be ascribed to the neglect of the variation of reduced mass with coordinate but reflects

1) a difference in the form of the normal coordinate for the two species and
2) a difference of zero-point averaging of the anharmonic interactions with the higher-frequency modes for the two species.

Again, this latter effect is of some importance. Referring to Table 4.10 and Eq. (4.7), we see that the vibrational dependence of the quartic term in the effective potential function is quite small, indeed within the quoted uncertainty. For cyclobutane, the reduced quartic potential constant is 26.15 ± 0.07 cm^{-1} for the ground state and 26.12 ± 0.07 cm^{-1} for the first excited state of the ν_{14} mode. On the other hand, the effect on the quadratic term is more noticeable, as expected from Eq. (4.7). For the ground state of ν_{14}, it is – 8.87 ± 0.03 cm^{-1} compared to – 8.76 ± 0.04 cm^{-1} for the excited state. From these data, we may conclude that the sign of the coefficient of the interaction term $Q_{14}^2Z^2$ is positive.

Table 4.9. Puckering structure on ν_{14} band of C_4D_8[27)]

Transition	ν_{obs}[a] (cm^{-1})	ν_{calc} (cm^{-1})	$\nu_{obs}-\nu_{calc}$ (cm^{-1})
1– 0	1083.6[b]	1084.73	–0.13[b]
2– 1	1239.3[b]	1240.12	–0.84[b]
3– 2	1083.6[b]	1082.99	+0.61
4– 3	1221.1	1220.82	+0.28
5– 4	1082.2[b]	1082.68	–0.48[b]
6– 5	–	1180.33	–
7– 6	1103.1	1104.04	–0.94
8– 7	1148.5	1148.51	–0.01
9– 8	1143.5	1143.86	–0.36
10– 9	1155.3	1155.52	–0.22
11–10	1161.6	1161.30	+0.30
12–11	1167.5	1167.12	+0.38
13–12	1172.8	1172.23	+0.57
14–13	1177.0	1176.94	+0.06
0– 1	1082.2[b]	1181.30	+0.90[b]
1– 2	–	927.50	–
2– 3	1082.2[b]	1082.91	–0.71
3– 4	943.8	942.61	1.19
4– 5	1079.2	1078.76	–0.84
5– 6	976.9	977.17	–0.27
6– 7	–	1052.25	–
7– 8	1010.4	1010.03	+0.37
8– 9	1016.0	1015.38	+0.62
9–10	1005.6	1005.92	–0.32
10–11	1001.3	1001.31	–0.01
11–12	996.6	996.60	0.00
12–13	992.3	992.43	–0.13
13–14	988.3	988.57	–0.27
14–15	984.6	984.96	–0.36
2– 4 }		{ 942.6	–
2– 5 }		{ 940.8	–
3– 4 }	943.8[b]	{ 944.4	–
3– 5 }		{ 942.7	–
7– 4	1208.9	1209.55	–0.65
5– 2	1222.7	1223.06	–0.36
3– 0 } 2– 1 }	1239.3[b]	1240.16	–0.86

[a] Data taken from Ref.[33)] [b] Blended or overlapped lines; not included in fit.

One other factor, which has a very minor effect on the barrier height, is the use of a Hamiltonian with a constant effective mass [Eqs. (3.22), (3.27)] as opposed to a Hamiltonian explicitly including the reduced-mass dependence on coordinate [e. g., Eq. (3.13)]. For the four-membered ring molecules treated with both types of Hamiltonians, the differences in barrier heights have been found to be 0–3 cm^{-1}, with the majority closer to 0 than to 3 cm^{-1}.

Table 4.10. Potential constants, band centers and barriers obtained for C_4H_8 and C_4D_8[a]

	A(cm^{-1})	−B	Barrier (cm^{-1})	ν_0	$\sigma(cm^{-1})$
		C_4H_8			
Ground state	26.153 ± 0.074	8.893 ± 0.034	514.8 ± 4.4	–	0.67
$\nu_{14} = 1$	26.117 ± 0.066	8.763 ± 0.039	501.4 ± 5.5	1454.6 ± 2.3	0.53
		$V'-V'' = 13.4 \pm 7.1$[b]			
		C_4D_8			
Ground state	18.59 ± 0.043	10.3768 ± 0.0027	500.6 ± 2.7	–	0.42
$\nu_{14} = 1$	18.561 ± 0.050	10.223 ± 0.059	485.0 ± 6.8	1084.7 ± 2.5	0.43
		$V'-V'' = 15.6 \pm 7.3$[b]			

[a] Errors cited are 3 standard deviations. Errors cited for the upper state constants are relative to those determined for the ground state.

[b] Error calculated from $\sigma_{\Delta V} = (\sigma'^2 + \sigma''^2)^{1/2}$.

b. Pseudo-Four-Membered Ring Molecules

(i) 2,5-Dihydrofuran. The far infrared spectrum of 2,5-dihydrofuran was reported by Ueda and Shimanouchi[12)] in 1967. They fitted the observed frequencies to a one-dimensional Hamiltonian similar to Eqs. (3.22), (3.27), with a positive quadratic coefficient indicating a planar molecule. In this case, although there are two out-of-plane degrees of freedom for the ring, one of the modes, the twisting about the C = C double bond, is relatively high in frequency. The other, essentially the motion of the oxygen normal to the plane of the other four ring atoms, is the low-frequency ring puckering. Figure 4.12 shows the far-infrared spectrum of 2,5-dihydrofuran obtained under higher resolution by Carriera and Lord[10)]. Clearly visible is a series of satellite transitions originating from the first excited state of the twisting mode. Effective one-dimensional potential functions were derived for each series by these workers.

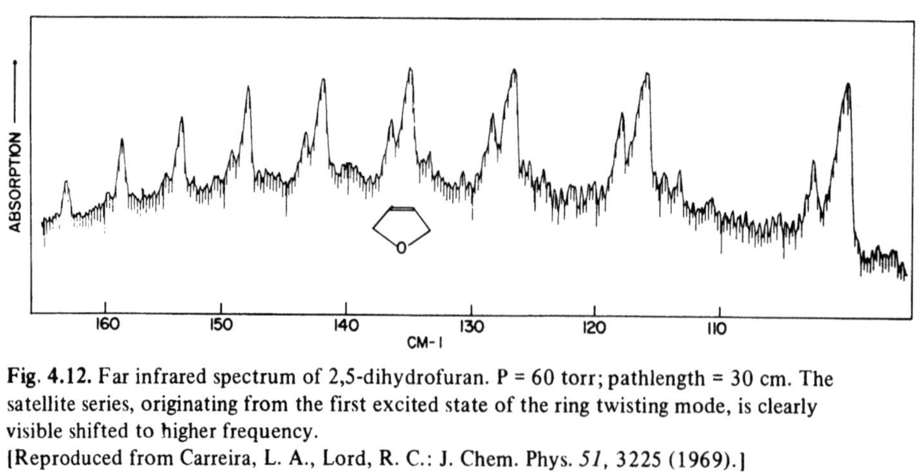

Fig. 4.12. Far infrared spectrum of 2,5-dihydrofuran. P = 60 torr; pathlength = 30 cm. The satellite series, originating from the first excited state of the ring twisting mode, is clearly visible shifted to higher frequency.
[Reproduced from Carreira, L. A., Lord, R. C.: J. Chem. Phys. *51*, 3225 (1969).]

Malloy[78)] considered the effect of including the coordinate dependence of the reduced mass in the Hamiltonian [Eq. (3.13)] for 2,5-dihydrofuran and other pseudo-four-membered rings. The Schrödinger equation used was

$$H = -\frac{\hbar^2}{2}\frac{d}{dx} g_{xx}(x) \frac{d}{dx} + ax^4 + bx^2 \tag{4.13}$$

where $g_{xx}(x)$ is the ring-puckering G-matrix element expressed as a function of x. Table 4.11 compares the fit to the data for a constant-reduced-mass Hamiltonian [Eq. (3.22)] with that for the above Hamiltonian in which $g_{xx}(x)$ is a least-squares polynomial for a semirigid model of the ring-puckering vibration. The fit is clearly much improved by the use of the latter. The importance of these terms for pseudo-four-membered rings compared to four-membered ring molecules has been discussed by Malloy and Lafferty[27)]. They found that the nature of the potential function does not change and the barrier heights derived from Eq. (4.13) are virtually identical with those from Eq. (3.22). Only the finer details are affected.

Carreira, Mills and Person[30)] fitted both ring-puckering series (Fig. 4.12) and ring-twisting data with a two-dimensional Hamiltonian explicitly including the interaction between these two modes [Eq. (3.33)]. Recently Malloy and Carreira[79)] have demonstrated the relationship between the effective one-dimensional potential functions which reproduce the ring-puckering series and the full two-dimensional potential function given in Eq. (3.33).

Table 4.11. Observed and calculated far infrared transition frequencies for 2,5-dihydrofuran[78)]

Transition	Obs[a] (cm^{-1})	Calc(I)[b] (cm^{-1})	Δ (cm^{-1})	Calc(II)[c] (cm^{-1})	Δ (cm^{-1})
0– 1	99.9	102.7	–2.8	100.4	–0.5
1– 2	116.2	115.4	+0.8	115.5	+0.7
2– 3	126.8	125.3	+1.3	126.3	+0.5
3– 4	135.2	133.5	+1.8	134.9	+0.3
4– 5	142.1	140.7	+1.4	142.2	–0.1
5– 6	148.1	147.2	+0.9	148.5	–0.4
6– 7	153.6	153.0	+0.6	154.0	–0.4
7– 8	158.5	158.4	+0.1	159.0	–0.5
8– 9	163.2	163.5	–0.3	163.5	–0.3
9–10	167.5	168.2	–0.7	167.5	0.0
10–11	171.3	172.6	–1.3	171.3	0.0
11–12	175.3	176.8	–1.5	174.6	+0.7
			[d] $\sigma^2 = 20.8$		[d] $\sigma^2 = 2.16$

$g_{xx}(x) = 0.1160 \times 10^{-1} - 0.3868 \times 10^{-1}x^2 - 0.4483x^4 + 2.087x^6$

[a] Ref.[10)].
[b] Least-squares fit with Eq. (3.22), the constant reduced mass Hamiltonian; all frequencies have unit weights.
[c] Least-squares fit with Eq. (4.13) for a semi-rigid model.
[d] σ^2 = sum of the squares of the deviations.

(ii) Cyclopentene. Cyclopentene is an example of a pseudo-four-membered ring molecule with a double-minimum potential function. The vibration-rotation interaction involving the v = 0 and 1 inversion doublet has been analyzed by Butcher and Costain[66]. A Hamiltonian described by Eqs. (4.9a, b) was used, with the a and b subscripts interchanged (i. e., the symmetry plane preserved as the molecule puckers is defined as the *b–c* plane). Using second-order perturbation theory, they derived the rotational constants for the two states, the vibration-rotation interaction constant and the 0–1 vibrational interval. An estimate of 250–400 cm^{-1} for the barrier was given. Laane and Lord[79a] reported the far-infrared spectrum shown in Fig. 4.13. Next to each strong transition is a satellite line originating from the first excited state of the ring-twisting. Laane and Lord fitted the infrared data and obtained effective one-dimensional double-minimum potential functions for the ground state series and the satellite series. The potential function, with a barrier of 232 cm^{-1}, for the ground state series is shown in Fig. 4.14. As for dihydrofuran, it was found that the fit to the data was improved by including the coordinate dependence of the reduced mass in the Hamiltonian [Eq. (3.33)]. In addition, the calculated 0–1 inversion splitting calculated from Eq. (3.33) was 0.89 cm^{-1} compared to 0.81 cm^{-1} from Eq. (3.22). The value determined from the analysis of the vibration-rotation interaction[66] was 0.91 cm^{-1}.

The Raman spectrum of gaseous cyclopentene[14] is shown in Fig. 2.4. The prominent features are the Q branch transitions ($\Delta v = 2$). This spectrum was one of the first Raman spectra of ring-puckering hot bands in the literature, and was reported independently by two groups[14, 80]. Combination and difference band progressions involving the ring-puckering vibration have been observed in the Raman and mid infrared spectra of cyclopentene[14, 49, 80].

Laane and co-workers have studied several deuterated analogs of cyclopentene, observing ring-puckering and ring-twisting transitions in the far-infrared and Raman spectra[81, 82]. Effective one-dimensional potential functions have been derived, and, as noted for four-membered rings, there is a minor isotopic dependence of the barriers derived from the effective one-dimensional potential functions. The barriers derived vary from 232 cm^{-1} for cyclopentene to 216 cm^{-1} for cyclopentene-d_8. The reduced-

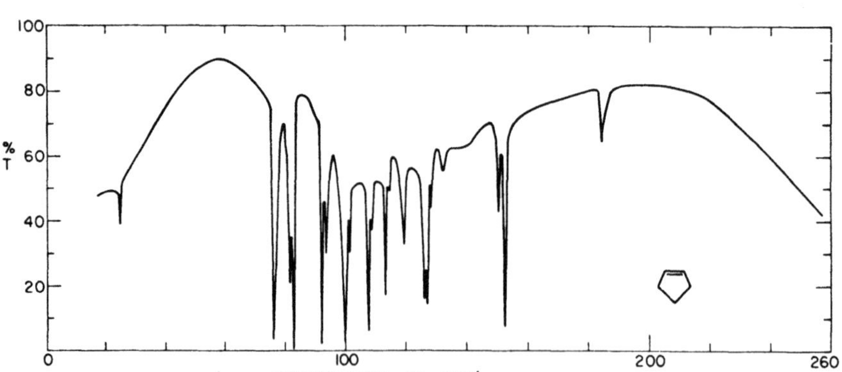

Fig. 4.13. Far infrared spectrum of cyclopentene. P = 115 torr; pathlength = 8 m. [Reproduced from Laane, J., Lord, R. C.: J. Chem. Phys. *47*, 4941 (1967).]

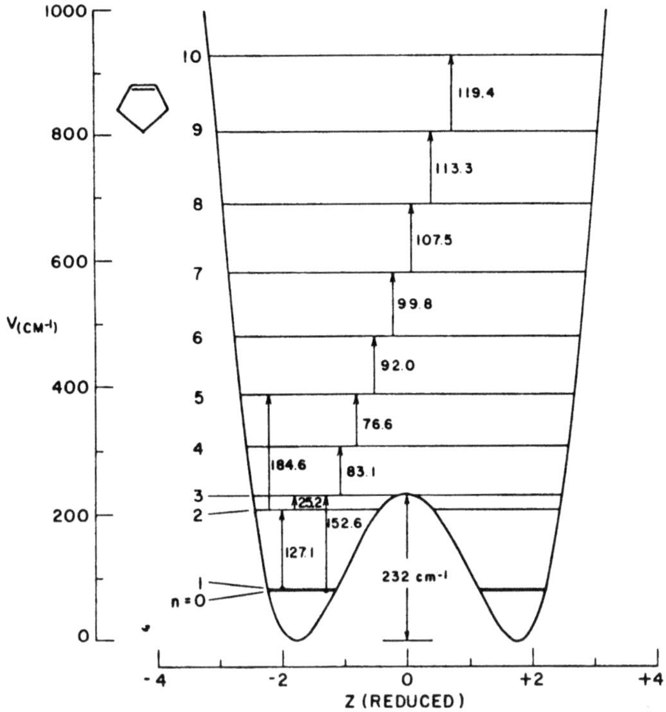

Fig. 4.14. Ring-puckering potential function for cyclopentene. [Reproduced from Laane, J., Lord, R. C.: J. Chem. Phys. *47*, 4941 (1967).]

mass dependence was included for several models. For a given isotopic species, this affected the determined barrier height by $\sim 2\ \mathrm{cm}^{-1}$. Calculations of two-dimensional potential surfaces for cyclopentene and various deuterated analogs using the Hamiltonian given in Eq. (3.33) are in progress[83].

(iii) 1,4-Dioxadiene. A six-membered ring molecule with two double bonds such as 1,4-dioxadiene may also behave as a pseudo-four-membered ring molecule. The B_{2u} ring-puckering vibration, which takes the molecule from the planar ring conformation to a boat form, is the mode of lowest frequency whose coordinate is defined in Fig. 4.13. The far infrared spectrum reported by Lord and Rounds[84] is shown in Fig. 4.15. The observed series of transitions increasing in frequency with a converging separation of adjacent transitions is the expected behavior for a planar molecule with a substantial quartic term in the potential function. Also evident in Fig. 4.15 is a series of ring-puckering transitions originating from the first excited

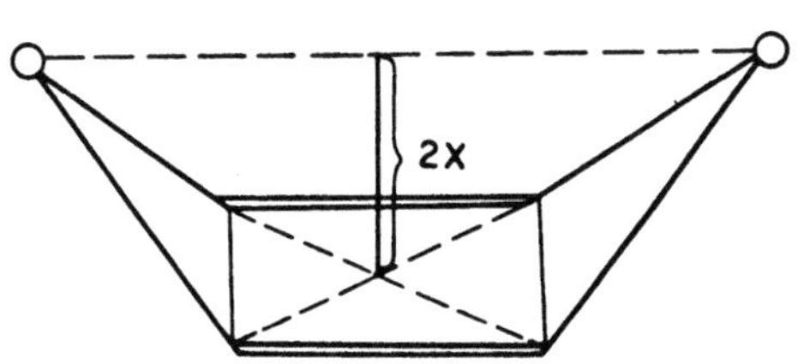

Fig. 4.15. Ring-puckering coordinate for 1,4-dioxadiene. [Reproduced from Lord, R. C., Rounds, T.C.: J. Chem. Phys. *58*, 4344 (1973).]

Table 4.12. Observed and calculated far infrared frequencies for 1,4-dioxadiene. Main series[84)]

Transition	Observed	Calculated[a]	Δ(Obs–calc)	Calculated[b]	Δ
0 → 1	66.82	68.26	−1.44	66.02	0.80
1 → 2	83.50	82.92	0.58	84.29	−0.79
2 → 3	93.75	92.80	0.95	94.49	−0.74
3 → 4	101.82	100.79	1.03	102.35	−0.53
4 → 5	108.55	107.60	0.95	108.78	−0.21
5 → 6	114.22	113.58	0.64	114.27	−0.04
6 → 7	119.15	118.97	0.18	118.95	0.20
7 → 8	123.72	123.89	−0.17	123.38	0.34
8 → 9	127.63	128.42	−0.79	126.91	0.72
9 → 10	130.81	132.59	−1.78	130.92	−0.11

a $\mu = 99.64$
$V(x) = 0.2949 \times 10^4 x^2 + 0.2930 \times 10^6 x^4$
$\sigma^2 = 9.5.$

b $g_{44}(x) = \frac{1}{\mu}(x) = 1.0036 \times 10^{-2} - 0.1389x^2 + 1.731x^4 - 8.79x^6$
$V(x) = 0.8251 \times 10^3 x^2 + 0.4702 \times 10^6 x^4$
$\sigma^2 = 2.8.$

state of the A_u ring-twisting mode. The shift in the ring-puckering frequencies on excitation of this mode is substantial, indicating a large x^2y^2 interaction term in the potential function. Despite the size of this term, it is possible to derive effective one-dimensional potential functions for each series. Table 4.12 shows the fit to the ground-state series with a constant-effective-mass Hamiltonian [Eq. (3.22)], and with a Hamiltonian that includes the coordinate dependence of the reduced mass [Eq. (4.13)]. Malloy and Carreira[79)] have recently derived a two-dimensional potential function like that in Eq. (3.33) and showed the relation between that function and effective one-dimensional functions derived by fitting the ring-puckering series separately.

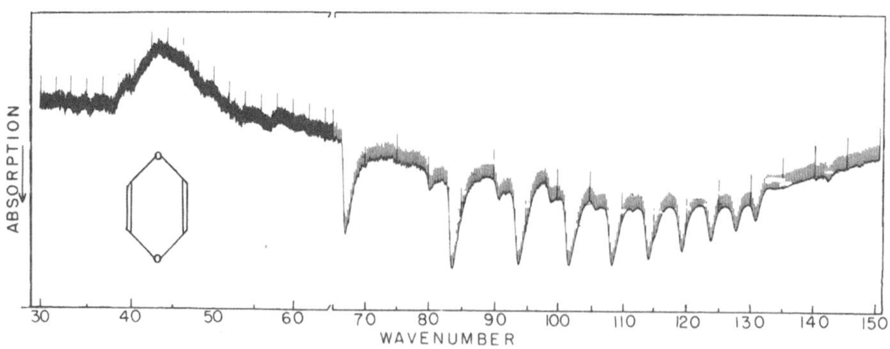

Fig. 4.16. Far infrared spectrum of 1,4-dioxadiene. The satellite series is observed shifted to higher frequency by a substantial amount. P = 25 torr; pathlength = 30 cm.
[Reproduced from Lord, R. C., Rounds, T. C.: J. Chem. Phys. *58*, 4344 (1973).]

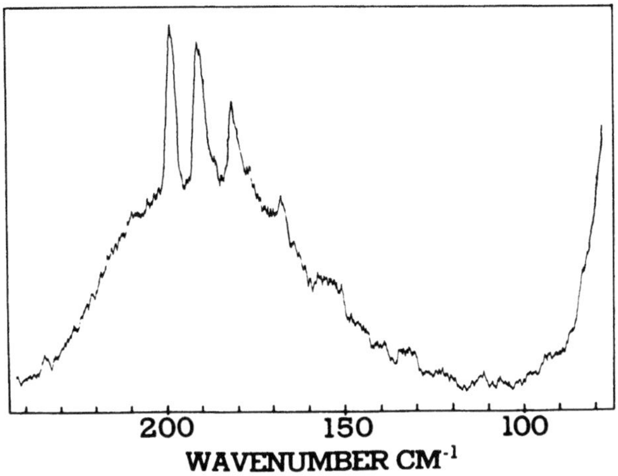

Fig. 4.17. Ring-puckering transitions in the Raman spectrum of gaseous 1,3-cyclohexadiene. [Reproduced from Carreira, L. A., Carter, R. O., Durig, J. R.: J. Chem. Phys. *59*, 813 (1973).]

(iv) 1,3-Cyclohexadiene. Another six-membered ring molecule with two endocyclic double bonds is 1,3-cyclohexadiene. In this case, the low-frequency mode is a ring twisting vibration that has a double-minimum potential. From his studies of the microwave spectrum of 1,3-cyclohexadiene, Butcher[85)] concluded that the equilibrium conformation has C_2 symmetry and that the angle between the $C_2{-}C_3$ and $C_5{-}C_6$ bonds is ~17.5°. The spectra observed were rigid-rotor spectra. The energy of the lowest excited state was estimated as $185 \pm 30\ cm^{-1}$ from relative intensity measurements of vibrational satellite lines in the microwave spectrum.

Carreira et al.[86)] studied the Raman spectrum of 1,3-cyclohexadiene vapor and observed a series of sharp Q branches probably due to $\Delta v = 2$ transitions of the ring-twisting vibration (Fig. 4.17). The double-minimum potential function derived by fitting the data with the two-parameter quartic-quadratic Hamiltonian is shown in Fig. 4.18. The barrier height is $1099 \pm 50\ cm^{-1}$, about twice as high as the highest

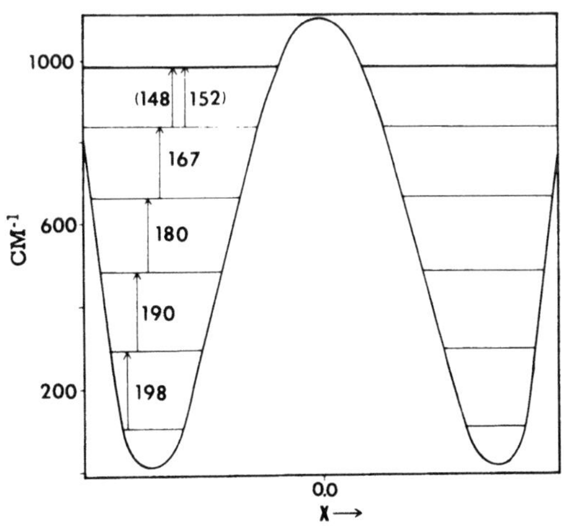

Fig. 4.18. Potential function determined for the ring-twisting vibration in 1,3-cyclohexadiene. [Reproduced from Carreira, L. A., Carter, R. O., Durig, J. R.: J. Chem. Phys. *59*, 813 (1973).]

barrier found for a saturated four-membered-ring molecule[13, 33]. The 0,1 levels are essentially degenerate as are the 2,3 levels, in agreement with the rigid-rotor microwave spectra observed for these pairs of levels. In the far infrared spectrum the type-b contours, along with the 4–5 cm^{-1} anharmonicity of the $\Delta v = 1$ transitions, wipe out the Q-branch maxima and permit no detailed assignment[87].

(v) 1,4-Cyclohexadiene. In striking contrast to the far-infrared spectrum of 1,4-dioxadiene (Fig. 4.16), the absorption of 1,4-cyclohexadiene due to its B_{2u} ring-puckering mode is confined to a narrow frequency range. Laane and Lord[88] observed a series of seven Q branches progressing to lower frequency from 108.4 cm^{-1} and spaced about 0.7 cm^{-1} apart. These were interpreted as the result of a one-dimensional Hamiltonian with a large quadratic term and a very small quartic term of sufficient magnitude to generate the slight separation observed. Probably the simplicity of the potential function is the result of co-operation rather than competition between ring strain and torsional interactions[88], both of which favor a planar molecule.

2 Asymmetric Systems

(i) Trimethylene Imine and 2,5-Dihydropyrrole. One of the first molecules for which an asymmetric potential function was determined was trimethylene imine. Due to the presence of the imino hydrogen the potential function is no longer symmetric. Figure 4.19 depicts the two non-equivalent puckered conformations.

The far-infrared spectra of trimethylene imine and trimethylene imine-N-*d* were reported by Carreira and Lord[89]. The Q branches are weak, superimposed on a background of rather strong overlapping P and R branches. The potential function determined for trimethylene imine, along with the energy levels and squared wavefunctions, is shown in Fig. 4.20. The left-well, right-well identifications of the states below the barrier are quite clear. For the fourth excited state, which is above the barrier, non-zero probability density occurs above both wells, but with greater amplitude above the left well. From v = 5 on up, it is difficult to make left-well, right-well identifications.

The gas-phase Raman spectra of trimethylene imine and trimethylene imine-N-*d* were studied later by Carreira et al.[90]. The transitions in the low-frequency region for trimethylene imine-N-*d* are shown in Fig. 4.21, and Table 4.13 summarizes the assignments of the transitions observed in the far infrared and Raman spectra. The

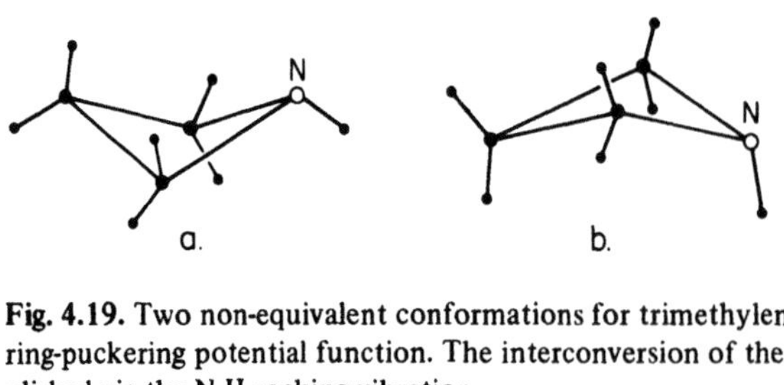

Fig. 4.19. Two non-equivalent conformations for trimethylene imine leading to an asymmetric ring-puckering potential function. The interconversion of these conformers may also be accomplished via the N-H rocking vibration.
[Reproduced from Carreira, L. A., Lord, R. C.: J. Chem. Phys. *51*, 2735 (1969).]

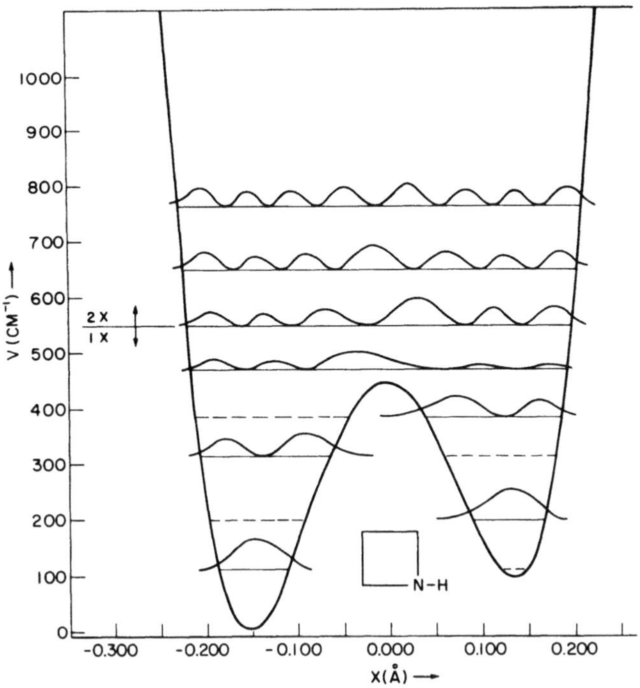

Fig. 4.20. Ring-puckering potential function for trimethylene imine. The squared wave functions illustrate the definite left well ⟷ right well identifications of the first four levels. [Reproduced from Carreira, L. A., Lord, R. C.: J. Chem. Phys. *51*, 2735 (1969).]

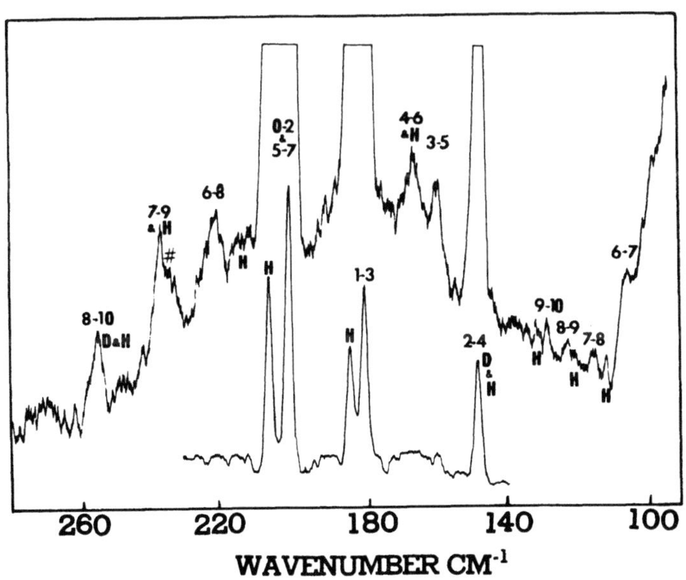

Fig. 4.21. Ring-puckering transitions in the Raman spectrum of trimethylene imine-N-*d*. [Reproduced from Carreira, L. A., Carter, R. O., Durig, J. R.: J. Chem. Phys. *57*, 3384 (1972).]

Table 4.13. Single and double jump ring-puckering transitions observed in the Raman effect for trimethylene imine[90)]

Transition	Observed Raman	Observed or inferred from far infrared	Δ
0 → 2	207.1	207.2	0.1
1 → 3	184.2[a]	184.6	0.4
2 → 4	148.6	149.3	0.7
3 → 5	166.1	b	–
4 → 6	184.2[a]	183.8	−0.4
5 → 7	216.9	216.4	−0.5
6 → 8	236.6[c]	237.0	0.4
7 → 9	255.8	255.4	−0.4
6 → 7	113.8?	113.7	−0.1
7 → 8	121.8?	123.3	1.5
8 → 9	133.9?	132.1	−1.8

[a] Both the 1 → 3 and the 4 → 6 transitions are expected at this frequency.
[b] The 3 → 4 transition was not observed in the infrared, since this transition originates from a level below the well to one above the well. In this case both the 3 → 4 and 3 → 5 transitions are calculated to be extremely weak in the infrared. The spacing of the 3 and 4 levels can be calculated from the observed 3 → 5 Raman transition and the observed 4 → 5 transition in the far infrared. This spacing is calculated to be 85.0 cm.
[c] This transition was difficult to measure due to a weak impurity band at 233.6 cm^{-1}.

observation of both infrared and Raman transitions, with different selection rules, serves to confirm the assignment and the double-minimum nature of the potential function. The barrier to interconversion of the two forms is 441 cm^{-1} above the deeper minimum. The energy difference between the two minima is 95 cm^{-1}. When the potential function is transformed to that of Eq. (3.32), the coefficients are found to have the values A = 29.84 cm^{-1}, B = 15.15, C = 7.73. Thus $9C^2 = 35.5B$, very close to the limit $9C^2 = 36B$ that corresponds to a symmetric double-minimum potential function (Table 3.1).

The fit to the data for trimethylene imine (rms deviation ~2.5 cm^{-1}) is not as good as has generally been obtained for molecules with symmetric potential functions. In this molecule there is a second pathway by which its two forms (Fig. 4.19) may be interconverted, namely, via the N-H inversion vibration. This vibration has the same symmetry properties as the ring-puckering. Consequently, harmonic, cubic and quartic cross terms are allowed in the potential. Neglect of these terms is doubtless one reason for the deviations observed when the data are fitted one-dimensionally.

A slight isotopic dependence of the effective one-dimensional potential function is found for trimethylene imine-N-*d*. However, this effect is of the same order as that observed for molecules with symmetric potential functions, and in this case is within the uncertainty of the potential functions. The barrier measured from the lowest well was 443 cm^{-1} and the energy difference between wells 90 cm^{-1}, compared to 441 cm^{-1} and 95 cm^{-1} for the parent compound.

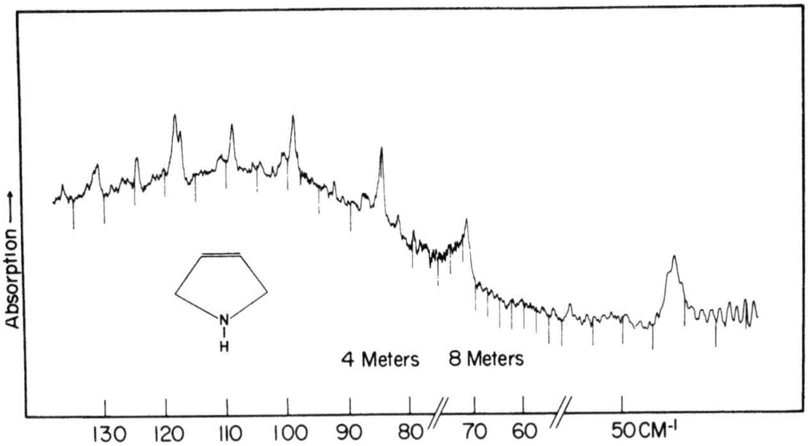

Fig. 4.22. Far infrared spectrum of 2,5-dihydropyrrole. P = 8 torr; pathlength = 4 m, 8 m as noted. [Reproduced from Carreira, L. A., Lord, R. C.: J. Chem. Phys. *51*, 2735 (1969).]

In spite of the factors mentioned in the previous paragraphs, the double-minimum nature of the potential functions for trimethylene imine and the N-d compound is well established. Comparison of the far infrared and Raman data for both compounds leaves no question as to the correct assignment.

2,5-Dihydropyrrole is a pseudo-four-membered ring molecule. Its structural relationship to trimethylene imine is the same as that of 2,5-dihydrofuran to trimethylene oxide. Again, both the parent and N-d compounds were studied. Figure 4.22 shows the far infrared spectrum of the light compound reported by Carreira and Lord[89]. The low-frequency Raman spectrum was reported later by Carreira et al.[90]. Transitions with both $\Delta v = 1$ and $\Delta v = 2$ were observed. Table 4.14 summarizes the

Table 4.14. Single and double jump ring-puckering transitions observed in the Raman effect for 2,5-dihydropyrrole[90]

Transition	Observed Raman	Observed or inferred from the far infrared	Δ
0 → 2	117.8[a]	118.1	0.3
1 → 3	155.7	156.3	0.6
2 → 4	184.1	183.8	−0.3
3 → 5	208.1	207.8	−0.3
4 → 6	227.3	226.2	−1.1
5 → 7	241.9	242.1	0.2
3 → 4	98.3	98.9	0.6
4 → 5	107.9	108.9	1.0
5 → 6	117.8[a]	117.3	−0.5
6 → 7	123.4	124.8	1.4
7 → 8	129.0	131.0	2.0
8 → 9	137.5	136.7	−0.8

[a] Both the 0 → 2 and 5 → 6 transitions are expected at this frequency.

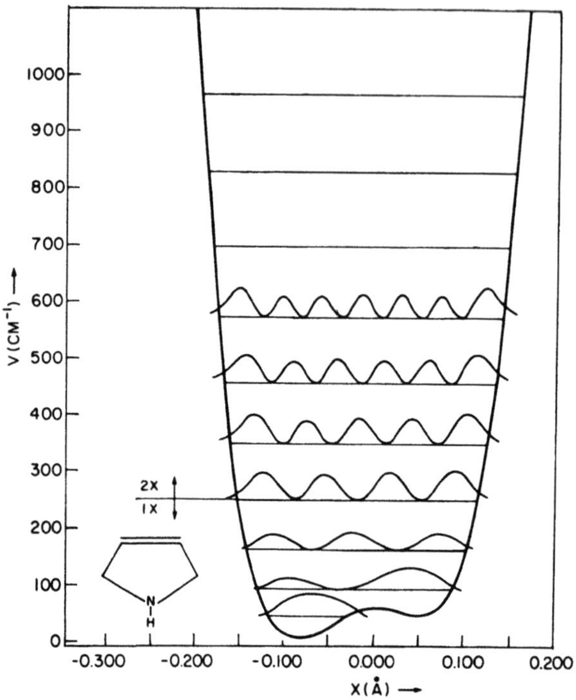

Fig. 4.23. Ring-puckering potential function for 2,5-dihydropyrrole. Only the lowest level has a definite left well character.
[Reproduced from Carreira, L. A., Carter, R. O., Durig, J. R.: J. Chem. Phys. *57*, 3384 (1972).]

assignments of the far-infrared and Raman transitions. Again, the different selection rules serve to confirm the assignments. Figure 4.23 gives the potential curve determined for the parent compound with the squared wave functions superimposed on the energy levels. There is a very small barrier between the two shallow minima and only one level, localized to the left well, occurs below the barrier. The v = 1 level has non-zero probability density above both the left and right wells, but with more over the right-well. From v = 2 on up, the probability density function is virtually symmetric. As in the case of trimethylene imine there is an isotopic dependence of the potential function, but the qualitative nature of the potential functions for 2,5-dihydropyrrole and 2,5-dihydropyrrole-N-*d* is the same.

(ii) Analogs of Bicyclo[3.1.0]hexane. Another type of pseudo-four-membered ring molecule with an asymmetric potential function is represented by bicyclo [3.1.0]-hexane and its analogs. The parent hydrocarbon (1) and three oxygen-containing analogs, 3-oxa-, 6-oxa- and 3,6-dioxabicyclo[3.1.0]hexane (2, 3, 4), have been studied. Far-infrared, Raman and microwave studies have been carried out[89, 91–96].

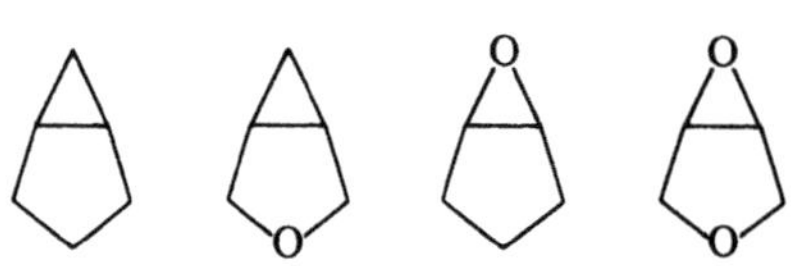

1 2 3 4

Three low-frequency out-of-plane ring vibrations may be characterized. The mode of highest frequency, a rocking of the atom (or CH_2 group) in position 6 in the ring, gives rise to a sharp intense Q branch in both the infrared and Raman spectra whose frequency ranges from 404 cm^{-1} for 6-oxabicyclo[3.1.0]hexane (cyclopentene oxide) to 365 cm^{-1} for 3.6-dioxabicyclo[3.1.0]hexane. The second mode is a twisting of the five-membered ring about the C_1-C_5 bond, observed as a type *b* band in the infrared and as a broad depolarized line in the Raman spectrum. Its frequency varies from 320 cm^{-1} for 6-oxabicyclo[3.1.0]hexane to 260 cm^{-1} for 3,6-dioxabicyclo[3.1.0]hexane. The ring-puckering mode yields the most information. This involves primarily the out-of-plane motion of the atom (or CH_2 group) in the 3 position and is highly anharmonic. The 0–1 transition frequencies range from 241 cm^{-1} for bicyclo[3.1.0]hexane to 195 cm^{-1} for 3,6-dioxabicyclo[3.1.0]hexane.

The combination of far infrared and Raman spectra proved especially effective in the study of these molecules, since those with intense spectra in the far infrared generally were found to have weak Raman spectra and vice-versa. Figure 4.24 presents the low-frequency Raman spectrum of the parent hydrocarbon (1)[96]. Transitions due to the three out-of-plane modes are evident in the figure, but only the ring puckering exhibits discernible hot-band structure. A series of six transitions with negative anharmonicity is observed. The first four of these transitions had been observed in the far-infrared spectrum under rather extreme conditions (80 torr pressure, 32 m path length)[92]. Table 4.15 lists the observed Raman and far infrared frequencies, and the fit to the Raman data by the three-parameter potential function of Eq. (3.32)[96]. Figure 4.25 shows this potential function, for which $9\,C^2 = 30.1$ B, so that it is a single-minimum function with two inflection points. The shape of the potential function is well determined up to the v = 6 level (~1250 cm^{-1} above v = 0), but above that it is extrapolated. The frequencies are predicted to reach a minimum with the 8–9 transition and then begin increasing in frequency due to the dominance of the quartic term [Eq. (3.32)] at large amplitudes. However, the Boltzmann population of these higher states is such that it was not possible to confirm this behavior.

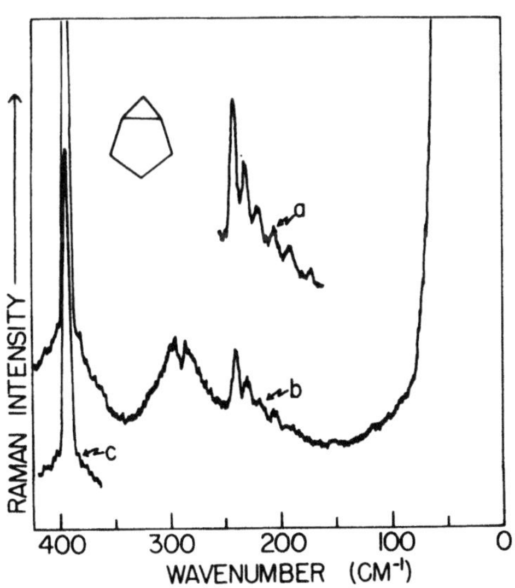

Fig. 4.24. Low-frequency region of the Raman spectrum of bicyclo[3.1.0]hexane. The Q branch at 390 cm^{-1} is assigned to rocking of the cyclopropane ring, the broad band at 290 cm^{-1} to the ring-twisting mode of the five-membered ring and the series of Q branches starting at 240 cm^{-1} to the ring-puckering vibration. [Reproduced from Lewis, J. D., Laane, J., Malloy, T. B., Jr.: J. Chem. Phys. *61*, 2342 (1974).]

Table 4.15. Ring-puckering transitions of bicyclo[3.1.0]hexane[96)]

Transition	Observed Raman frequency $(cm^{-1})^a$	Calculated frequency $(cm^{-1})^b$	Δ (cm^{-1})	Raman intensity	Calculated intensity[c]	Depolarization ratio	Observed IR frequency $(cm^{-1})^d$
0 → 1	241.0	240.9	0.1	1.0	1.0	<0.1	240.7
1 → 2	230.4	230.4	0.0	0.6	0.6	<0.2	230.8
2 → 3	218.8	218.7	0.1	0.3	0.3	<0.2	218.4
3 → 4	204.6	205.5	−0.9	0.2	0.13	–	204.3
4 → 5	191.0	190.2	0.8	0.2	0.06	–	–
5 → 6	172.0	172.2	−0.2	0.1	0.03	–	–
6 → 7	–	151.1	–	–	0.01	–	–

a ± 0.5 cm^{-1} except for 172.0 cm^{-1}, ± 1.0.
b $V(cm^{-1}) = 24.70\,(Z^4 + 25.72\,Z^2 + 9.28\,Z^3)$.
c From $\langle v'|x|v\rangle$; for upper transitions $\langle v'|x^2|v\rangle$ becomes significant.
d Ref.[92)].

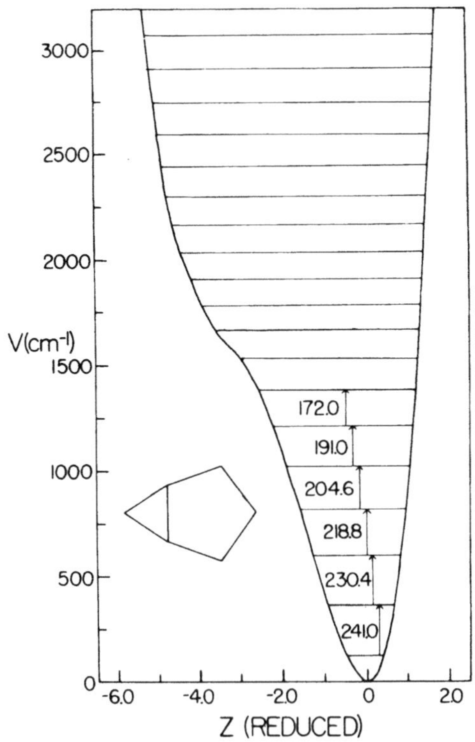

Fig. 4.25. Single-minimum asymmetric potential function for the ring-puckering vibration in bicyclo[3.1.0]hexane. [Reproduced from Lewis, J. D., Laane, J., Malloy, T. B., Jr.: J. Chem. Phys. *61*, 2342 (1974).]

3,6-dioxabicyclo[3.1.0]hexane proved to be a more favorable case[92]. The ring-puckering frequencies are somewhat lower so that the Boltzmann factors permitted the observation of more transitions. In addition, the large dipole moment changes resulting from the presence of the two oxygen atoms produced a very intense spectrum in the far infrared (Fig. 4.26). The assignments of the nine transitions and the fit to the data are given in Table 4.16. Figure 4.27 shows the potential function, for which $9C^2 = 28.1\,B$. Again, a single-minimum potential function with two inflection points has resulted from fitting the data, but in this case, data have been obtained for levels which are well above the second inflection point and the shape of the potential has been well characterized. Figure 4.27 shows that the cubic term in the potential contributes substantial negative anharmonicity for the first few transitions. The quartic term then causes the frequencies to reach a minimum with the 5–6 transition and then increase in frequency for the 6–7, 7–8 and 8–9 transitions. Because transitions up into the quartic-dominated region have been observed, there is no question that the ring-puckering potential function for 3,6-dioxabicyclo[3.1.0] hexane has only a single minimum.

The seven observed Q-branch transitions for 3-oxabicyclo[3.1.0]hexane were fitted[92] to a similar function with $9C^2 = 26.8\,B$. The 6–7 transition was the last observed Q-branch. The 7–8 and 8–9 transitions were predicted to be the lowest frequencies, being almost identical, and succeeding transitions were indicated to increase in frequency. On the other hand, the potential function reported by Carreira and Lord[89] for 6-oxabicyclo[3.1.0]hexane (cyclopentene oxide) had a second, very

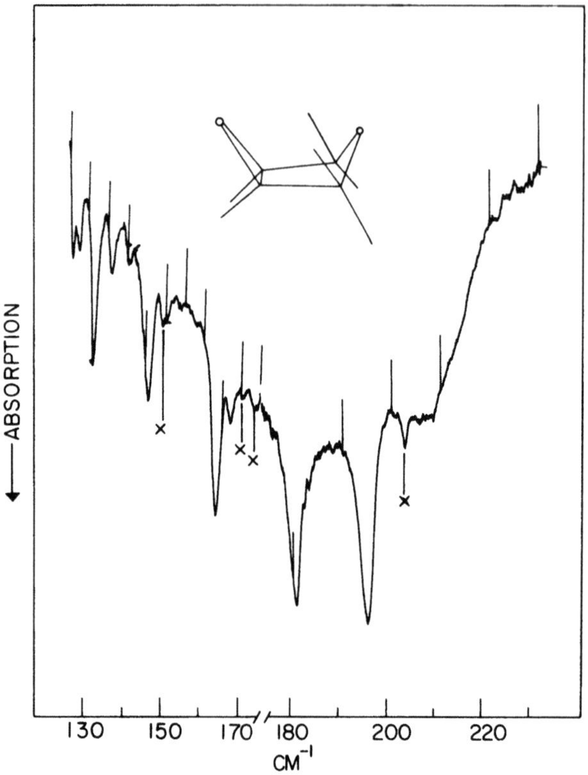

Fig. 4.26. Far infrared spectrum of 3,6-dioxabicyclo[3.1.0]hexane. P ~1 torr; pathlength = 20 m. x denotes a water peak.
[Reproduced from Lord, R. C., Malloy, T. B., Jr.: J. Mol. Spectroscopy *46*, 358 (1973).]

Table 4.16. Observed and calculated ring-puckering transitions for 3,6-dioxabicyclo[3.1.0]hexane[a]

Transition	Observed (cm^{-1})	Calculated[b] (cm^{-1})	Δ (cm^{-1})	Calculated intensity
0 → 1	195.1	196.0	−0.9	1.00
1 → 2	180.7	180.7	0.0	0.70
2 → 3	164.2	163.5	+0.7	0.38
3 → 4	146.1	145.7	+0.4	0.20
4 → 5	131.9	131.7	+0.2	0.11
5 → 6	126.1	126.9	−0.8	0.07
6 → 7	128.0	129.4	−1.4	0.04
7 → 8	136.1	134.4	+1.7	0.03
8 → 9	140.4	139.9	+0.5	0.02

a Ref.[92]).
b $V(cm^{-1}) = 25.25\ (Z^4 + 17.23Z^2 + 7.33Z^3)$.

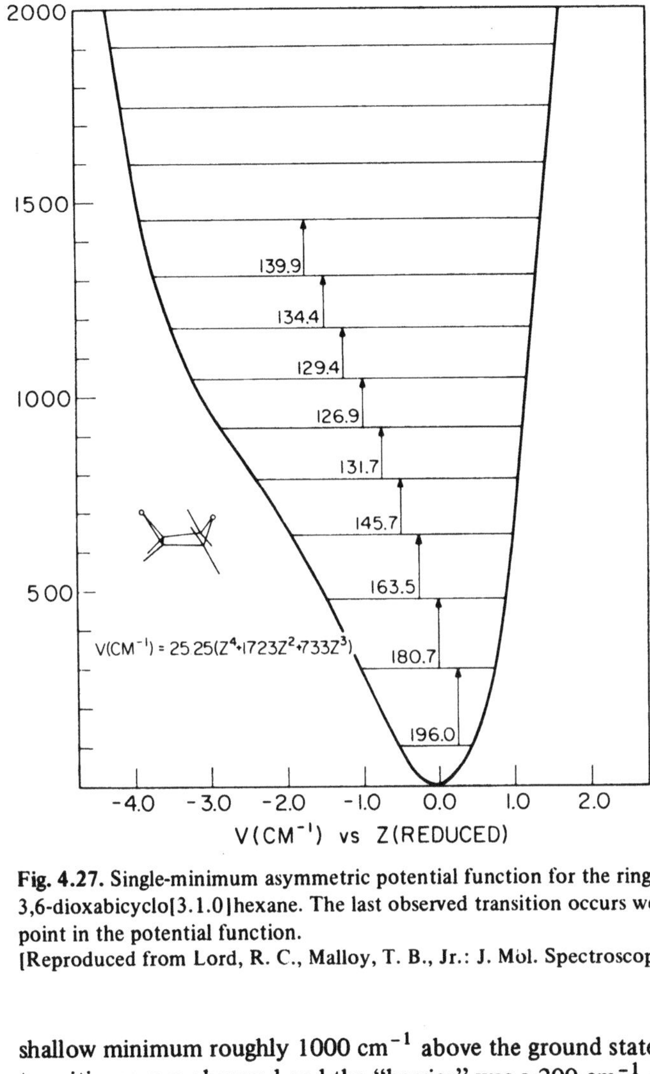

Fig. 4.27. Single-minimum asymmetric potential function for the ring-puckering vibration of 3,6-dioxabicyclo[3.1.0]hexane. The last observed transition occurs well above the inflection point in the potential function.
[Reproduced from Lord, R. C., Malloy, T. B., Jr.: J. Mol. Spectroscopy *46*, 358 (1973).]

shallow minimum roughly 1000 cm^{-1} above the ground state. However, only five transitions were observed and the "barrier" was ~200 cm^{-1} above the n = 5 level. For this potential function, $9C^2 = 32.4B$, very close to the dividing line, $9C^2 = 32B$, between double-minimum and single minimum potential functions. Carreira and Lord weighted more heavily the lower, more intense transitions in their fit of the five transitions for cyclopentene oxide, obtaining a very shallow second minimum. When their data were refitted, with equal weights assigned to each transition, a single-minimum potential function resulted. The important point is that with only five transitions *the potential functions were well within one standard deviation of each other.* Thus one must regard as questionable any feature in a potential function that lies much above the last observed energy level.

The determination of the potential function in reduced coordinates from fitting far-infrared or Raman data does not yield information on the identity of the stable conformer. The microwave spectrum of 3,6-dioxabicyclo[3.1.0]hexane was reported by Creswell and Lafferty[93], who showed from the ground-state rotational constants and the dipole-moment components that the boat conformer is the stable form. Similar studies have been reported for 6-oxabicyclo[3.1.0]hexane[91].

B. Molecules Treated by Two-Dimensional Hamiltonians

1 Pure Pseudorotation

(i) Cyclopentane. Durig and Wertz[7] reported the pseudorotational combination and difference transitions superimposed on a CH_2 deformation mode in the mid infrared spectrum of cyclopentane. They interpreted these with the help of Eq. (3.38). Carreira et al. employed the perturbation formulas Eqs. (3.39) to (3.41) and three radial transitions observed in the vapor-phase Raman spectrum[97] to make a provisional estimate of the parameters in a two-dimensional potential function. With these as a starting point they carried out a least-squares fitting of all the data by numerical solution of the two-dimensional problem in polar coordinates. This allowed them to estimate the barrier to planarity in cyclopentane as 1824 ± 50 cm^{-1}, quite close to the value obtained by Pitzer and Donath[98] from thermodynamic data.

2 Hindered Pseudorotation

(i) Tetrahydrofuran and 1,3-Dioxolane. Originally the angular transitions in the far infrared spectrum of tetrahydrofuran and 1,3-dioxolane were interpreted on the basis of pure pseudorotaion using Eq. (3.38) and higher quantum-number transitions, from levels above the small barriers[6, 99]. The microwave spectrum of tetrahydrofuran revealed a complicated energy pattern for low quantum numbers[38]. Rotational

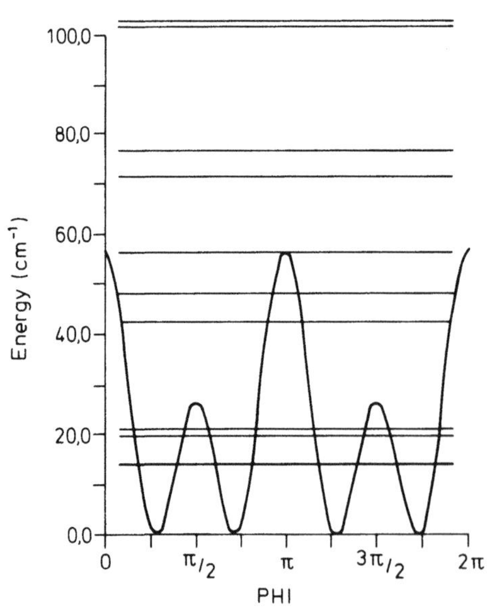

Fig. 4.28. Angular potential function hindering pseudorotation in tetrahydrofuran. [Reproduced from Engerholm, G. G., Luntz, A. C., Gwinn, W. D., Harris, D. O.: J. Chem. Phys. *50*, 2446 (1969).]

constants in different vibrational states were determined, as well as small energy splittings between the 0–1 levels and the 2–3 levels, for which the Schrödinger Eq. (3.44) was used. The rotational constants were not expressed in power series, but in a trigonometric series appropriate to a periodic coordinate

$$\beta_{v\phi} = \sum_k \beta^{(k)} \langle v_\phi | \cos k\phi | v_\phi \rangle \tag{4.14}$$

where β represents the A, B, or C rotational constant. The splittings and variations in rotational constants were used to determine the following potential function

$$V(\phi) \text{ in cm}^{-1} = -15(1 - \cos 2\phi) - 20(1 - \cos 4\phi) \tag{4.15}$$

The value of the pseudorotational constant, 3.25 cm^{-1}, was taken from the earlier far-infrared study[6]. Figure 4.28 depicts the potential function hindering pseudorotation. It is seen that the largest barrier encountered in one cycle is ~55 cm^{-1}, indicating that the approximation that the pseudorotation barrier be much less than the barrier to planarity has been met in this case.

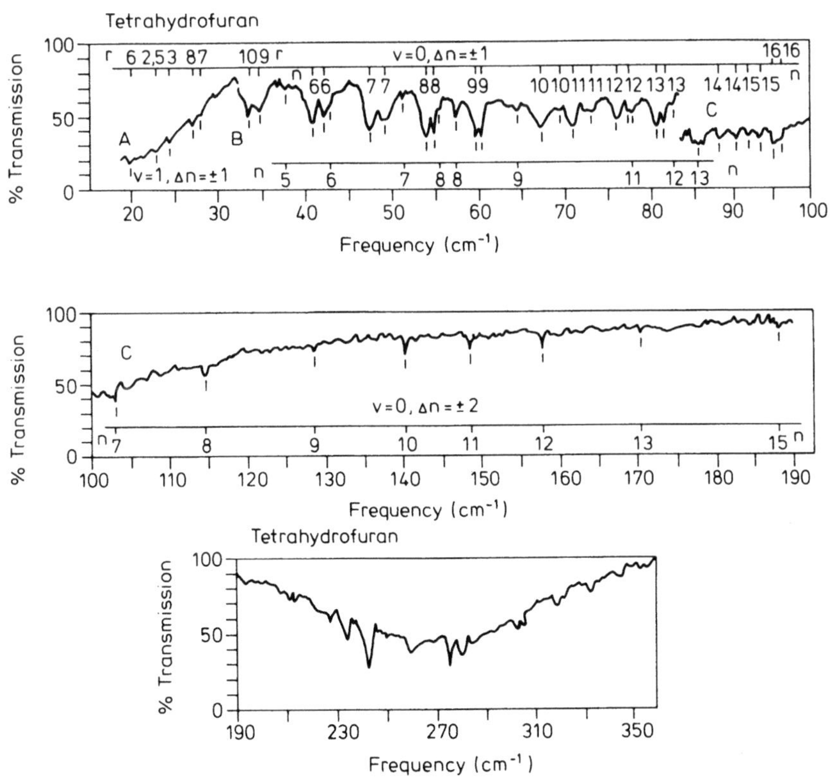

Fig. 4.29. Far infrared spectrum of tetrahydrofuran. Path length = 1 m. (A) P = 20 torr, (B) P = 20 Torr, (C) P = 93 torr. P = 43 torr in the region above 190. [Reproduced from Greenhouse, J. A., Strauss, H. L.: J. Chem. Phys. *50*, 124 (1969).]

Under conditions of higher resolution, the far infrared spectrum of tetrahydrofuran was found to be quite complex (Fig. 4.29) with a number of bands split by Coriolis interaction. Assignments and analysis of the band origins led to a potential function

$$V(\phi) \text{ in } cm^{-1} = -13.5(1 - \cos 2\phi) - 20(1 - \cos 4\phi) \tag{4.16}$$

with a pseudorotational constant of 3.19 cm^{-1}. The features in the radial-band region (ca. 270 cm^{-1}) remained unassigned. Subsequent treatment of the far infrared and microwave splitting data with a two-dimensional Hamiltonian in polar coordinates [Eq. (3.43)] and also including coordinate-dependent mass terms in the kinetic energy, led to tentative assignments of some of these features[100)]. The calculated spectrum is quite complex and shows that many of the features are due to several overlapped transitions. Recently, Sont and Wieser[101)] have made assignments of several features in the Raman spectrum of tetrahydrofuran near 270 cm^{-1}.

3 Coupled Bending and Twisting Vibrations in Cyclopentanone

In some cases, separation of variables in polar coordinates is not appropriate. Cyclopentanone is a good example of such a molecule. Although it is possible to fit the main features of the far infrared spectrum with a periodic potential function, the "barrier" derived has no meaning. A rather thorough study of the far infrared spectrum of cyclopentanone by Ikeda and Lord[102)] led to the determination of the potential surface shown in Fig. 4.30 by fitting the spectrum two-dimensionally. These workers reported the far infrared spectra of the parent compound and its α-d_4, β-d_4 and d_8 derivatives shown in Fig. 4.31. The data for the parent compound were fitted by least squares to a two-dimensional Hamiltonian similar to Eq. (4.13), but including variable reduced-mass terms through quadratic expansion coefficients in the kinetic energy operator. The stable conformation is the twisted (C_2) ring conformation, as had been shown by a microwave study[103)]. Not only is the barrier to planarity comparable to the barrier to pseudorotation, but also the minimum-energy path for interconversion of the equivalent C_2 conformers passes through the planar conformation with a relatively low barrier of 750 cm^{-1}.

Examination of Fig. 4.30 indicates that the probability density for the levels involved in the observed transitions should be localized about the stable twisted conformation. In order to obtain a basis set for the variation calculation which satisfies this requirement a slightly different approach was taken[102)]. First, the two-dimensional potential was presumed to be given by

$$V(x, y) = a_1x^4 + b_1x^2 + a_2y^4 + b_2y^2 + a_{12}x^2y^2 \tag{4.17}$$

where x is the bending coordinate and y the twisting coordinate. For cyclopentanone, it turned out that b_2 is negative while each of the other potential constants is positive.

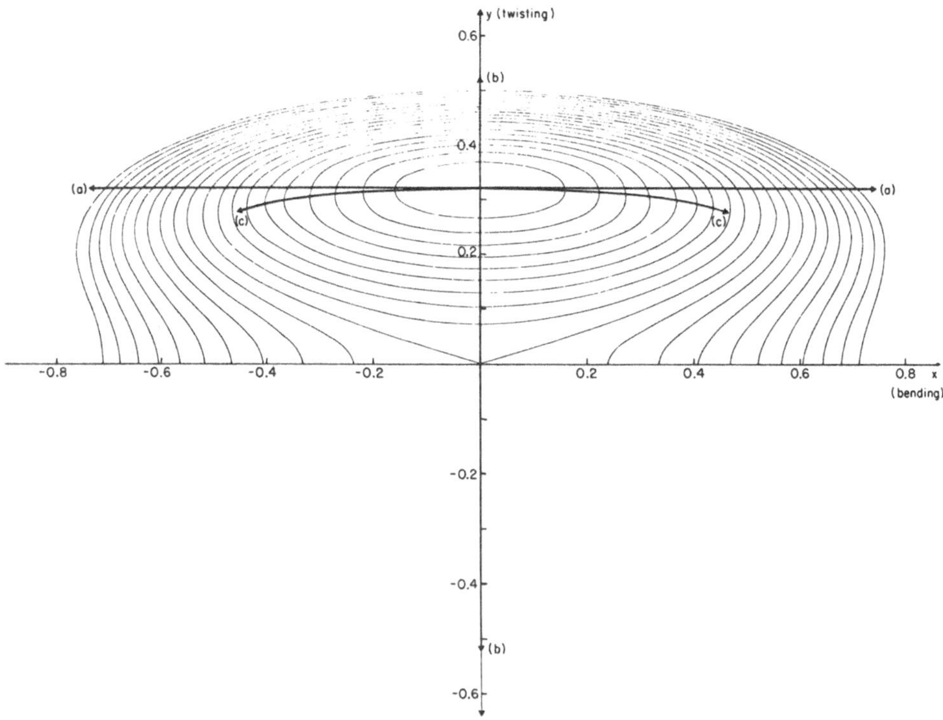

Fig. 4.30. Potential energy contour diagram for cyclopentanone. The contour in the third and fourth quadrants is the mirror image of that in the first two quadrants. The interconversion of the two equivalent C_2 conformers occurs via the planar conformation.
[Reproduced from Ikeda, T., Lord, R. C.: J. Chem. Phys. *56*, 4450 (1972).]

An approximate separation of variables was performed, yielding the following effective potential functions

$$V(y) = a_{12}y^4 + b_2y^2 \qquad (4.18)$$

and

$$V(x) = a_1x^4 + (b_1 - b_2a_{12}/2a_2)x^2 \qquad (4.19)$$

The origin in Eq. (4.19) has been translated to correspond to the minimum along the twisting coordinate and appropriate modifications made to the kinetic energy operator. The potential function then has a double minimum in the twisting coordinate and a single minimum in the bending coordinate. The resulting Schödinger equations were solved, as described earlier, with harmonic-oscillator basis functions. The solutions to these equations were then used to form a direct-product basis in which to expand the full two-dimensional Hamiltonian. This Hamiltonian matrix factored into four blocks denoted as *ee, eo, oe,* and *oo* depending on the even or odd character of the functions used to form the product basis. This procedure was then included in an iterative least-squares adjustment of the potential constants.

Figure 4.32 indicates the calculated spectra for the various deuterated analogs of cyclopentanone. As found earlier with cyclobutane and other ring molecules, it was necessary to mix motion of the methylene hydrogens (deuteriums) with the ring vibrations in order to reproduce the observed isotopic shifts. It may be seen from the figure that deuteration at the α-position has a much greater effect than deuteration at the β-position.

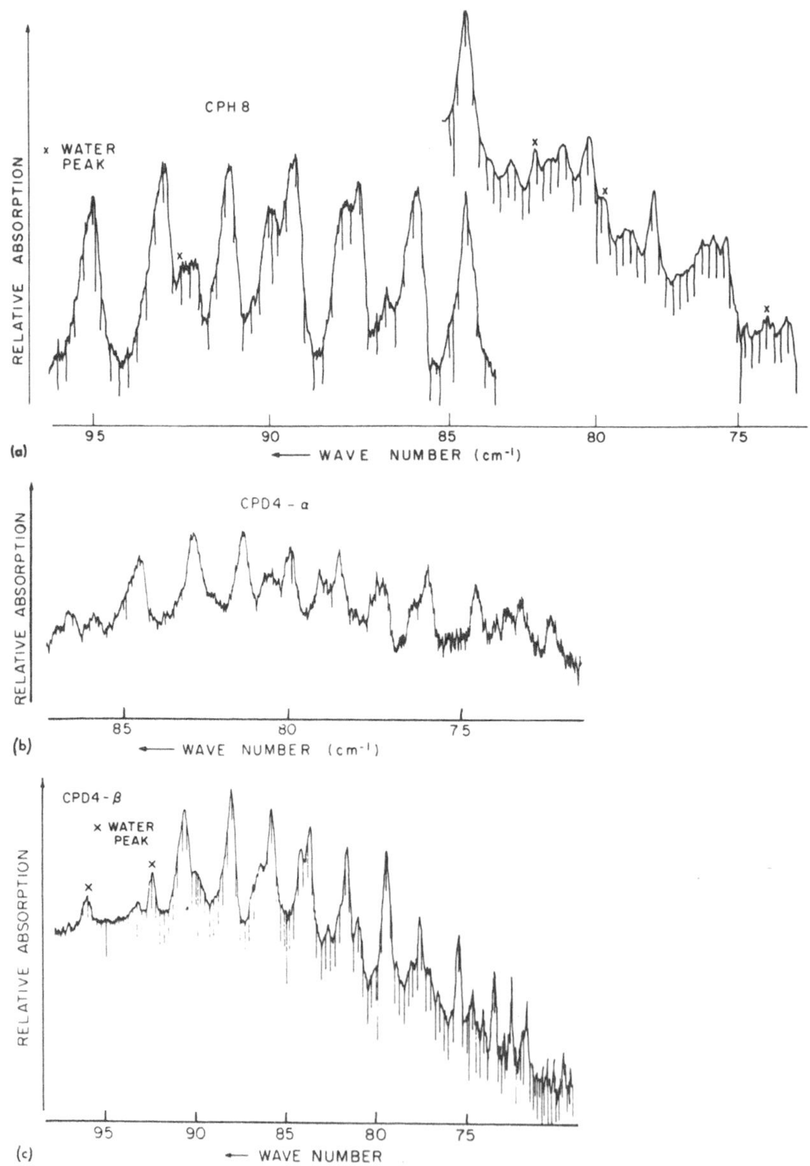

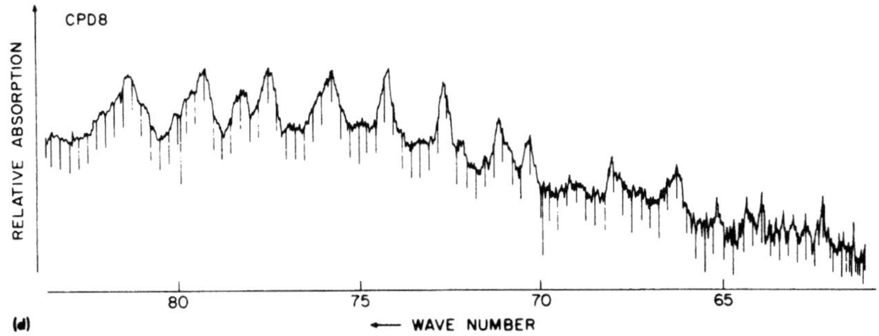

Fig. 4.31. Far-infrared spectra in the bending region for cyclopentanone, α, α, α', α'-cyclopentanone-d_4, β, β, β', β'-cyclopentanone-d_4 and cyclopentanone-d_8.
[Ikeda, T., Lord, R. C.: J. Chem. Phys. *56*, 4450 (1972).]

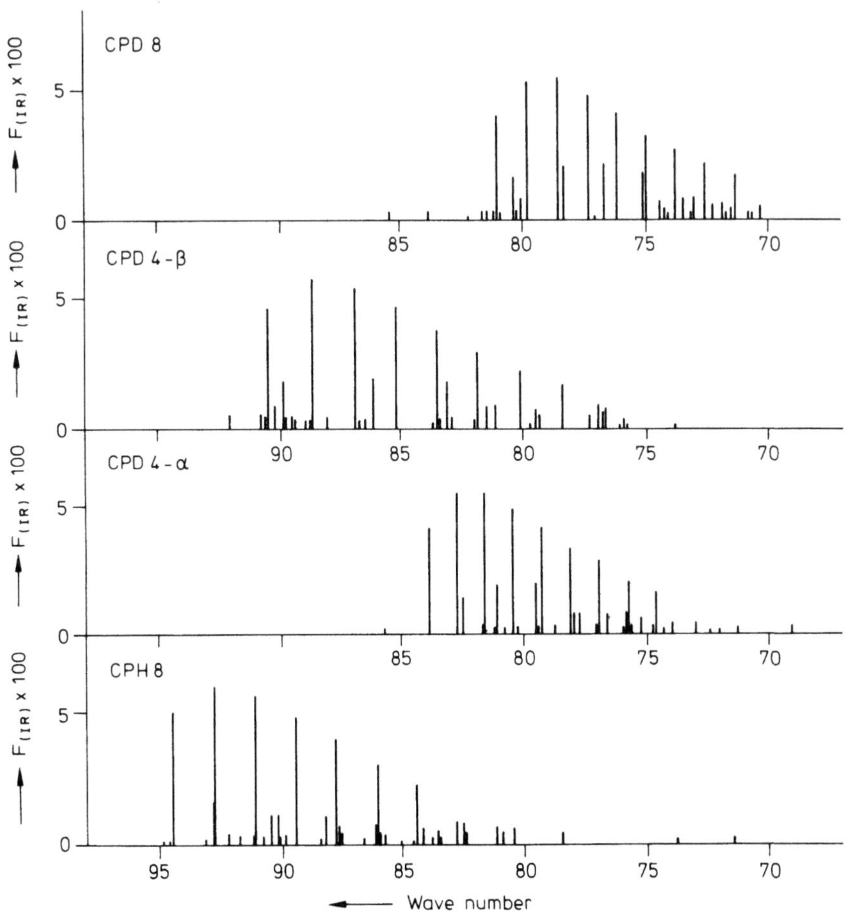

Fig. 4.32. Calculated spectra of cyclopentanone-d_0, α-d_4, β-d_4 and d_8.
[Reproduced from Ikeda, T., Lord, R. C.: J. Chem. Phys. *56*, 4450 (1972).]

4 Pseudo-Five-Membered Ring Molecules

Just as a five-membered ring molecule with an endocyclic double bond may be considered as a pseudo-four-membered ring, a six-membered ring molecule with an endocyclic double bond may be considered as a pseudo-five-membered ring molecule. Although a six-membered ring has three out-of-plane skeletal vibrations, in the case of analogs of cyclohexene one of these is mainly a twisting about the double bond, which is somewhat higher in frequency than the other two and essentially harmonic. The other modes, illustrated for 1,4-dioxene in Fig. 4.33, are a bending, β and twisting, τ. These two low-frequency modes are expected to be highly coupled.

(i) 1,4-Dioxene. The far infrared spectrum of dioxene[104] is shown in Fig. 4.34. The high-frequency region near 300 cm^{-1} involves transitions among "twisting" states, giving rise to overlapped type-b band contours. On the other hand, the "bending" transitions have type-c band contours and extensive hot-band structure is observed just below 200 cm^{-1}. At long path lengths, a series of bands arising from difference transitions between twisting and bending states is observed near 100 cm^{-1}.

Figure 4.35 reproduces the Raman spectrum of dioxene in the region corresponding to the twisting vibration[105]. The Q-branch transitions are prominent because the twisting mode is a totally symmetric vibration. Combining all of these data, one can build up a detailed energy-level pattern with many internal checks on the self consistency of the assignment[104, 105]. Figure 4.36 shows this pattern and the assignments of the transitions. The data have been fitted by least squares by the procedure described for cyclopentanone. The observed and calculated frequencies are given in Table 4.17.

In Fig. 4.37 is shown a potential energy contour map determined for 1,4-dioxene that has absolute minima at the two equivalent twisted (C_2) conformations; relative minima occur at the bent (C_s) conformations. The planar conformation corresponds to a maximum. Two cross sections of this surface, one along the twisting coordinate through the origin and the second along the minimum-energy path between the two equivalent C_2 conformations, are plotted in Fig. 4.38. The barriers involved are considerably higher for this molecule than for the four- or five-membered rings considered earlier. It should also be emphasized that the barriers are higher than the energy of the highest observed levels, that is, they have been determined by extrapolation and the uncertainties are correspondingly greater than for molecules with lower barriers. The position of the relative minimum corresponding to the C_s con-

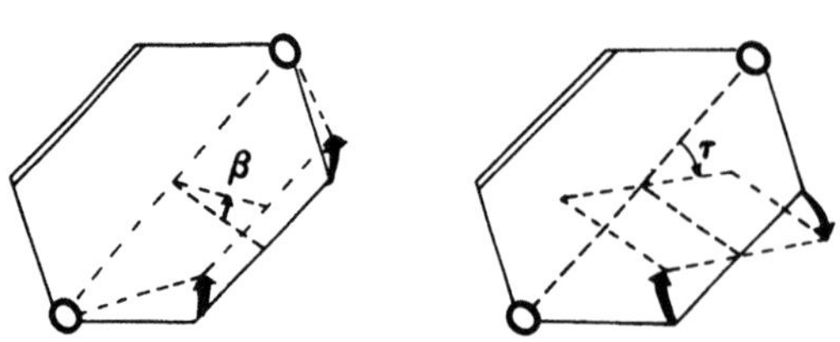

Fig. 4.33. Definitions of ring bending (β) and ring twisting (τ) coordinates for dioxene. [Reproduced from Lord, R. C., Rounds, T. C., Ueda, T.: J. Chem. Phys. 57, 2572 (1972).]

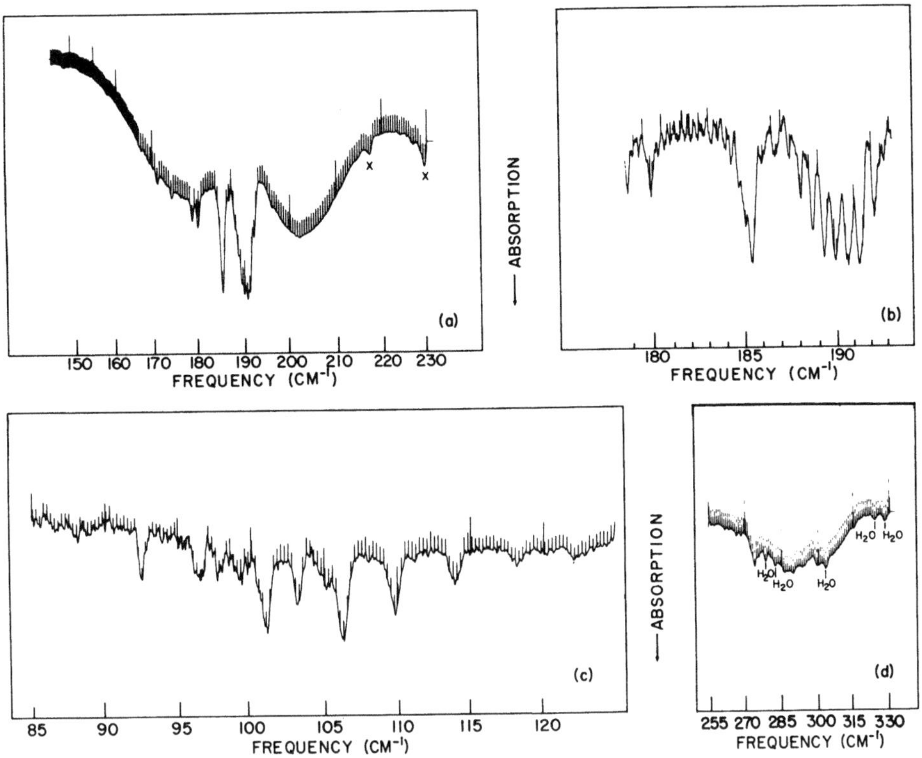

Fig. 4.34. Far infrared spectrum of dioxene. (A) Ring bending transitions P = 35 torr; path length = 30 cm. (B) High-resolution scan of the bending region showing resolved rotational structure. P = 35 torr: path length = 30 cm. (C) Difference band region, P = 10 torr; path length = 8 m. (D) Type-*b* bands in the ring-twisting region. P = 10 torr; path length = 4 m. [Reproduced from Lord, R. C., Rounds T. C., Ueda, T.: J. Chem. Phys. *57,* 2572 (1972).]

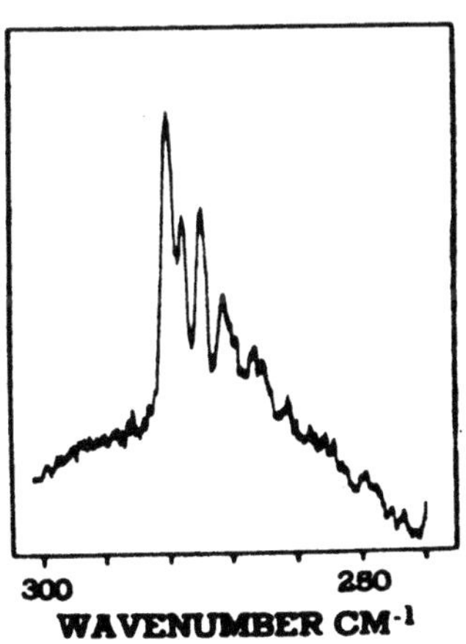

Fig. 4.35. Ring-twisting hot bands in the Raman spectrum of dioxene.
[Reproduced from Durig, J. R., Carter, R. O., Carreira, L. A.: J. Chem. Phys. *60,* 3098 (1974).]

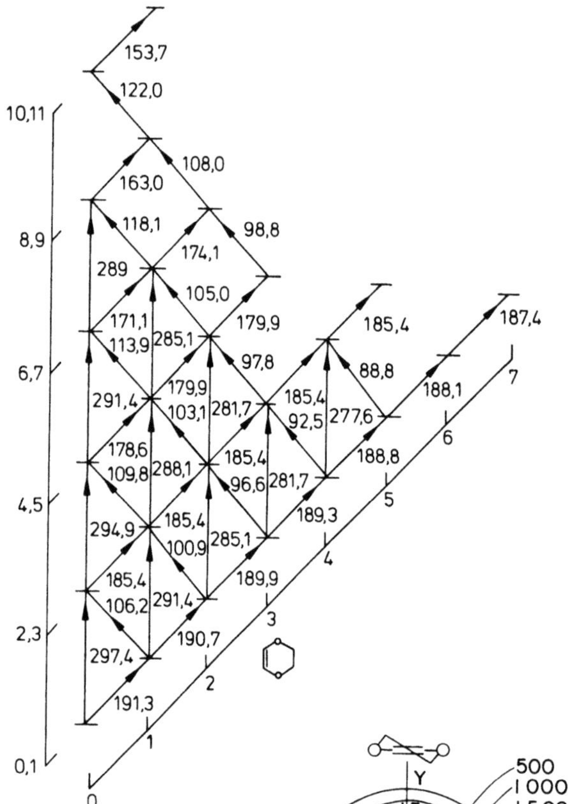

Fig. 4.36. Pattern of observed transitions between the energy levels for the twisting and bending vibrations in dioxene. Many of the checks on the self-consistency of the assignments are provided by the difference band in the far-infrared spectrum (Fig. 4.34c).
[Reproduced from Durig, J. R., Carter, R. O., Carreira, L. A.: J. Chem. Phys. *60*, 3098 (1974).]

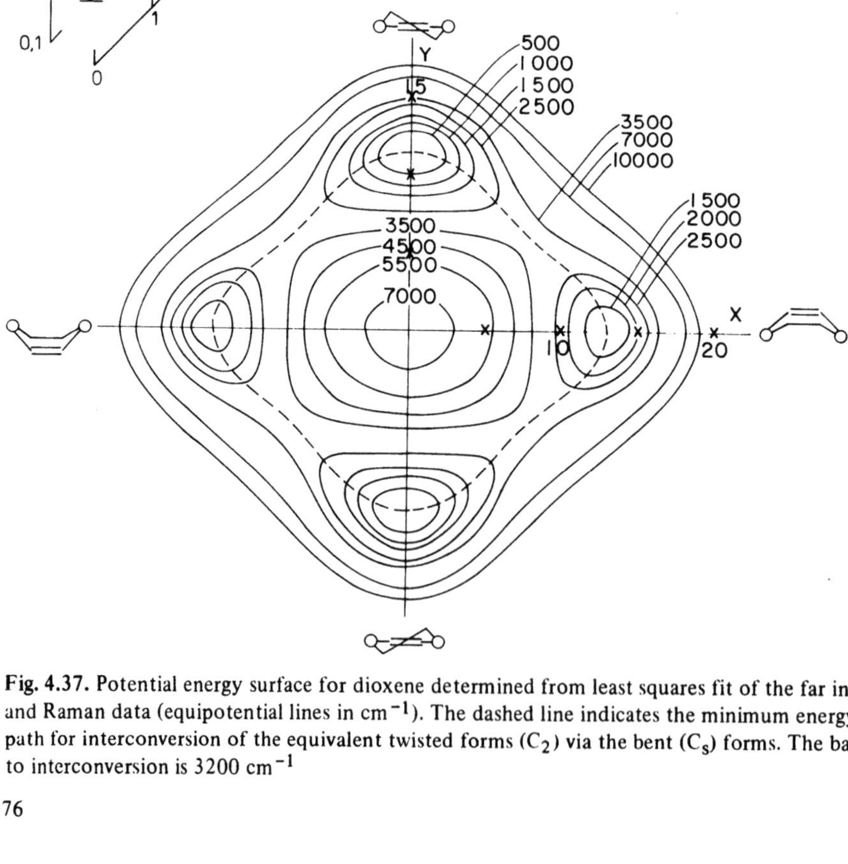

Fig. 4.37. Potential energy surface for dioxene determined from least squares fit of the far infrared and Raman data (equipotential lines in cm^{-1}). The dashed line indicates the minimum energy path for interconversion of the equivalent twisted forms (C_2) via the bent (C_s) forms. The barrier to interconversion is 3200 cm^{-1}

Table 4.17. Observed and calculated frequencies (cm^{-1}) for 1,4-dioxene

Obs	Assignment $(v_t, v_b) \rightarrow (v'_t, v'_b)$	Calc	Δ
Twisting band region[a]			
297.4	(0,0) → (2,0)	297.3	+0.1
294.9	(2,0) → (4,0)	294.7	+0.2
291.4	(4,0) → (6,0)	291.9	−0.5
289.0[b]	(6,0) → (8,0)	288.8	+0.2
291.4	(0,1) → (2,1)	291.7	−0.3
288.1	(2,1) → (4,1)	288.6	−0.5
285.1	(4,1) → (6,1)	285.1	0.0
285.1	(0,2) → (2,2)	286.4	−1.3
281.7	(2,2) → (4,2)	283.0	−1.3
281.7	(0,3) → (2,3)	281.3	+0.4
277.6	(0,4) → (2,4)	276.3	+1.3
Bending band region[b]			
191.3	(0,0) → (0,1)	190.4	+0.9
190.7	(0,1) → (0,2)	190.3	+0.4
189.9	(0,2) → (0,3)	190.1	−0.2
189.3	(0,3) → (0,4)	189.8	−0.5
188.8	(0,4) → (0,5)	189.4	−0.6
188.1	(0,5) → (0,6)	189.0	−0.9
187.4	(0,6) → (0,7)	188.5	−1.1
	(2,0) → (2,1)	184.8	+0.6
	(2,1) → (2,2)	185.0	+0.4
185.4	(2,2) → (2,3)	185.0	+0.4
	(2,3) → (2,4)	184.8	+0.6
	(2,4) → (2,5)	184.5	+0.9
178.6	(4,0) → (4,1)	178.7	−0.1
179.9	(4,1) → (4,2)	179.4	+0.5
179.9	(4,2) → (4,3)	179.6	+0.3
171.1	(6,0) → (6,1)	171.9	−0.8
174.1	(6,1) → (6,2)	173.5	+0.6
163.0	(8,0) → (8,1)	164.3	−1.3
Difference band region[b]			
106.2	(0,1) → (2,0)	106.9	−0.7
109.8	(2,1) → (4,0)	109.8	0.0
113.9	(4,1) → (6,0)	113.2	+0.7
118.1	(6,1) → (8,0)	116.7	+1.4
100.9	(0,2) → (2,1)	101.4	−0.5
103.1	(2,2) → (4,1)	103.6	−0.5
105.0	(4,2) → (6,1)	105.7	−0.7
108.0	(6,2) → (8,1)	107.6	+0.4
96.6	(0,3) → (2,2)	96.3	+0.3
97.8	(2,3) → (4,2)	98.0	−0.2
98.8	(4,3) → (6,2)	99.6	−0.8
92.5	(0,4) → (2,3)	91.5	+1.0
88.8	(0,5) → (2,4)	86.9	+1.9

[a] From the Raman spectrum (Ref.[105]). [b] From the far-infrared spectrum (Ref.[104]).

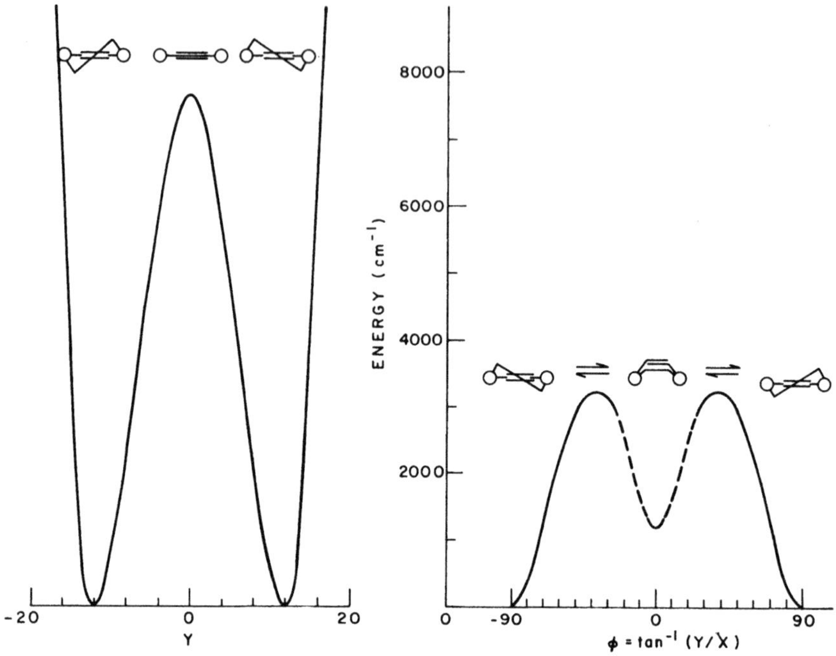

Fig. 4.38. Cross sections of the potential surface in Fig. 4.37. The figure on the left is the potential function along the twisting axis. The potential function on the right follows the dashed line in Fig. 4.37 and represents the minimum-energy path. The portion of the curve corresponding to the C_s conformer is dashed, indicating the large uncertainty in the position of the second minimum

formation is shown in Fig. 4.38 as a dotted line to imply that the shape of the curve in this region and the energy of the C_s minimum are not at all well determined. The dispersion of the barrier to planarity determined from the least squares fit to the data is ± 750 cm^{-1}. The dispersion of the barrier to interconversion of the two equivalent C_2 forms is ± 250 cm^{-1}. On the other hand, the dispersion of the depth of the second minimum is of the order of this depth.

The barriers to interconversion determined for 1,4-dioxene and similar unsaturated six-membered rings are approaching the upper limit of those which can be determined reliably from vibrational data alone at temperatures near 300 K. On the other hand, they overlap the lower limit of those which may be determined from the study of the temperature dependence of the nmr spectra. A subsequent study of the nmr spectrum of dioxene in solution as a function of temperature led to an independent determination of the barrier to interconversion[106]. The agreement with that determined from fitting the vibrational data was within 1 kcal/mole. This was quite satisfactory in as much as the two experiments were performed in different phases. In addition, the angle between the C–C and C=C bonds was estimated[107] from the vicinal coupling constants as 26.8°, in reasonable agreement with the value of 29.9° calculated from a microwave study[108].

A list of small ring molecules investigated up to the middle of 1978 is summarized in Table 4.18 under the headings of Section IV. (See pp. 80–90).

V Prospects for Further Studies

Further research in the field of low-frequency vibrations in ring molecules may be expected to proceed along three lines:

1) Extension of studies like those treated in this article to additional small-ring systems, including isotopic derivatives. Such studies will provide new information about the molecules studied as well as further tests of the theoretical treatments discussed in Section III.

2) Extension of the investigations to larger rings. The low-frequency vibrational modes of saturated six-membered and larger ring molecules have already been studied by Strauss and coworkers (see, for example, Refs.[109–112]). Their approach has been to develop a molecular force field based on vibrational data and known molecular geometries for a series of molecules rather than to determine the potential surfaces for individual molecules experimentally from the spectra of one or two low-frequency vibrations. Clearly both the theoretical and experimental problems associated with the larger rings are difficult. However, new experimental techniques (see below), especially those with the capability of much higher spectral resolution, seem likely to provide the data needed to treat these more complicated systems. A review of the present status of the vibrational spectroscopy of medium-sized rings is now available[113].

3) New developments in experimental methods. New techniques for high-resolution spectroscopy in far infrared absorption and the Raman effect are likely to improve both the accuracy and the scope of studies of ring systems. It was noted in Section II that the increasing application of Fourier-transform infrared spectroscopy to the investigation of ring molecules has already resulted in markedly improved accuracy of the molecular parameters derived therefrom. Much further improvement can be expected from the methods of laser spectroscopy. The use of tunable diode lasers, for example, gives promise of spectral resolutions of the order of 10^{-4} cm^{-1} in the mid and far infrared[114]. Clearly the quality of the data affordable by such resolving power will necessitate corresponding improvement in the theoretical treatment.

Since the high-resolution Raman spectra of ring compounds are of necessity obtained from samples in the vapor phase, improvement in the resolution is dependent on an increase in the intensity of the scattered radiation. A promising technique for this purpose is that of coherent anti-Stokes Raman scattering (CARS), which is many orders of magnitude more intense than the normal Raman effect. Recent developments in the technique[115] suggest that CARS could produce spectra of sufficient intensity and line sharpness to enable the resolution of lines separated by 0.001–0.01 cm^{-1}. While this range is not so impressive as the potential resolution of the tunable lasers, it is still three orders of magnitude better than present work. In any case the combination of much greater spectral detail with extended theory should permit the successful study of larger molecules and improved accuracy of the potential surfaces of smaller rings.

Table 4.18 A. Four-membered ring molecules with symmetric potential functions

Formula	Structure	Name	Techniques	Comments	Ref.
$C_2F_4S_2$	CF_2SCF_2S	1,1,3,3-Tetrafluordithietane	FIR	Planar, harmonic	116)
$C_2H_8Si_2$	$CH_2SiH_2CH_2SiH_2$	1,3-Disilacyclobutane	FIR, MIR, R, MR	Double minimum potential function. Barrier = 87 cm^{-1}. Isotopic Species	117)
$C_3H_4F_2O$	$OCH_2CF_2CH_2$	3,3-Difluorooxetane	FIR, MW	Potential function determined. Planar	118, 119)
$C_3H_4O_2$	CH_2CCH_2O ‖ O	Oxetanone-3	FIR, MW	Potential function determined. Planar. Bending occurs mostly about C—C diagonal	10, 15, 120)
$C_3H_4O_2$	CH_2COCH_2 ‖ O	Oxetanone-2	MW	Planar. No potential function determined	121)
C_3H_4OS	CH_2CCH_2S ‖ O	Thietanone-3	FIR, MW	Potential function determined. Planar. Bends primarily about C---C diagonal	43, 44)
C_3H_6O	$CH_2CH_2CH_2O$	Trimethylene oxide	FIR, MIR, MW, R, MR	Potential function determined. Barrier = 15.3 cm^{-1}, less than the zero point energy. Extensive studies of isotopic species. Extensive studies of vibration rotation interaction	5, 9, 32) 45–59)

C_3H_6S	$CH_2CH_2CH_2S$	Trimethylene sulfide	FIR, MIR, MW, R	Double minimum potential function. Barrier = 274 cm^{-1}. Extensive study of deuterated species showed a minor isotopic dependence of the barriers (ca. 10 cm^{-1}) determined from the effective one-dimensional constant-mass Hamiltonian	18, 21) 63–69)
C_3H_6Se	$CH_2CH_2SeCH_2$	Trimethylene selenide	FIR, MIR, MW	Double minimum potential function. Barrier = 373 cm^{-1}	75, 76)
C_3H_8Si	$CH_2CH_2SiH_2CH_2$	Silacyclobutane	FIR, MIR, R MR, MW	Double minimum potential function. Barrier = 440 cm^{-1}. SiD_2 species studied	71–74)
$C_4H_4F_4$	$CH_2CF_2CF_2CH_2$	1,1,2,2-Tetrafluorocyclobutane	MW, FIR	Non planar based on MW data. No potential function determined	122, 123)
$C_4H_4O_2$	$OC(=CH_2)CH_2C(=O)$	Diketene	FIR, MW	Planar, harmonic	10, 124)
$C_4H_6F_2$	$CH_2CH_2CF_2CH_2$	1,1-Difluorocyclobutane	MW	Double minimum potential function determined from the variation of rotational constants and from analysis of the non-rigid rotor spectra in the v = 0, 1 and 2,3 inversion states. Barrier = 241 cm^{-1}	125)
C_4H_6O	$CH_2CH_2C(=O)CH_2$	Cyclobutanone	FIR, MW, MIR	Potential function determined. Barrier = 7.6 cm^{-1}, less than the zero point energy. Several isotopic species studied	16, 49, 63) 64, 126)

Table 4.18 A. (continued)

Formula	Structure	Name	Techniques	Comments	Ref.
C_4H_6O	$\overline{OCH_2CCH_2}$ (=CH_2)	3-Methyleneoxetane	MW, FIR	Potential function determined. Planar. Bends mostly about C---C diagonal	127, 128)
C_4H_8	$\overline{CH_2CH_2CH_2CH_2}$	Cyclobutane	MIR, R	Double minimum potential function. Barrier = 515 cm^{-1}. C_4H_8 studied. Minor isotopic dependence (15 cm^{-1}) of the barrier derived from effective one-dimensional potential functions. Evidence for mixing of CH_2 rocking with ring puckering	13, 27) 33, 77)
C_5H_8	$\overline{CH_2CH_2CCH2}$ (=CH_2)	Methylenecyclobutane	MW, MIR, R	Double minimum potential function. Barrier = 140 cm^{-1}. Isotopic species studies (MIR)	17, 21) 65, 70)
$C_5H_{10}O$	$\overline{CH_2CCH_2O}$ (=$(CH_3)_2$)	3,3-Dimethyloxetane	FIR	Double minimum potential function determined. Barrier = 46 cm^{-1}	129)
C_6H_8	$\overline{CCH_2CH_2C}$ (=CH_2, =CH_2)	1,2-Dimethylenecyclobutane	MW	Planar based on smooth variation of rotational constants with vibrational state. Potential function not determined	130)
B_2H_6	$\overline{BH_2HBH_2H}$	Diborane	IR, R	Potential function determined. Planar. Isotopic species studied	131, 132)
B_2H_6S	$\overline{BH_2SHBH_2H}$	μ-Mercaptodiborane	IR	Potential function determined. Planar. Isotopic species studied	133)
B_2H_7N	$\overline{BH_2NH_2BH_2H}$	μ-Aminodiborane	IR	Potential function determined. Planar Isotopic species studied	134)

Table 4.18B. Pseudo-four-membered ring molecules with symmetric potential functions

Formula	Structure	Name	Techniques	Comments	Ref.
$C_3H_2O_3$	CH=CHOC(=O)O (ring)	Vinylene carbonate	MW	Planar based on variation of rotational constants. No potential function determined. Isotopic species studied	135, 136)
C_3H_5NO	$CH_2OCH{=}NCH_2$ (ring)	2-Oxazoline	MIR, MW	Potential function determined from MIR band progressions and MW data. Planar	137)
$C_3H_6N_2$	$CH_2N{=}NCH_2CH_2$ (ring)	1-Pyrazoline	FIR	Double minimum potential function determined. Barrier = 113 cm^{-1}	78, 138)
$C_4H_4O_2$	CH=CHOCH=CHO (ring)	1,4-Dioxadiene	FIR	Potential function determined. Planar. Evidence for interaction of ring puckering with ring twisting. (See Table 4.18D)	84)
$C_4H_6F_2Si$	$CH{=}CHCH_2SiF_2CH_2$ (ring)	1,1-Difluorosilacyclo-pent-3-ene	MW	Planar based on variation of rotational constants. Potential function not determined	139)
C_4H_6O	$CH{=}CHCH_2OCH_2$ (ring)	2,5-Dihydrofuran	FIR, MIR, R	Potential function determined. Planar. Evidence for interaction of ring-puckering with ring-twisting. (See Table 4.18D)	10, 12, 30) 78–80)
C_4H_6O	$CH{=}CHOCH_2CH_2$ (ring)	2,3-Dihydrofuran	FIR, MIR, MW, R	Double minimum potential function determined. Barrier = 83 cm^{-1}	48, 78) 140–142)

Table 4.18B. (continued)

Formula	Structure	Name	Techniques	Comments	Ref.
C_4H_6S	$\overline{CH=CHCH_2SCH_2}$	2,3-Dihydrothiophene	MIR, R	Double minimum potential function determined. Barrier = 325 cm^{-1}	48, 144, 143)
C_4H_8Si	$\overline{CH=CHCH_2SiH_2CH_2}$	Silacyclopent-3-ene	FIR, MIR, R	Potential function determined. Planar. Essentially a pure quartic oscillator. SiD_2 species studied	72, 73) 78, 144)
C_4H_8Si	$\overline{CH=CHSiH_2CH_2CH_2}$	Silacyclopent-2-ene	FIR, MIR	Potential function determined Planar	73, 145)
C_5H_6O	$\overline{CH=CHCH_2C(=O)CH_2}$	3-Cyclopentene-1-one	FIR, MW	Potential function determined. Planar. Isotopic species in MW	146–148)
C_5H_6O	$\overline{CH=CHC(=O)CH_2CH_2}$	2-Cyclopentene-1-one	FIR, MW	Potential function determined. Planar	149, 150)
C_5H_8	$\overline{CH=CHCH_2CH_2CH_2}$	Cyclopentene	FIR, MIR, R	Double minimum potential function determined. Barrier = 232 cm^{-1}. Evidence for interaction between ring-puckering and ring-twisting (See Table 4.IV)	14, 20, 21, 48) 49, 66, 78–82)
C_5H_8O	$\overline{CH=C(CH_3)OCH_2CH_2}$	2-Methyl 4,5-Dihydrofuran	FIR	Double minimum potential function. Barrier = 98 cm^{-1}	10)

C_6H_8	$CH{=}CHCH_2CH{=}CHCH_2$ (ring)	1,4-Cyclohexadiene	FIR, R	Planar	86, 88)
C_6H_8	$CH{=}CHCH{=}CHCH_2CH_2$ (ring)	1,3-Cyclohexadiene	MW, R	Symmetry = C_2 from MW. Potential function in twisting coordinate is double minimum. Barrier = 1099 cm^{-1}	85, 86)

Table 4.18C. Ring molecules with asymmetric one-dimensional potential functions

Formula	Structure	Name	Techniques	Comments	Ref.
C_3H_6OS	$CH_2CH_2SCH_2$ (ring), S=O	Trimethylene sulfoxide	MW	Puckered ring. =O equatorial. Potential function not determined	151)
C_3H_7N	$CH_2CH_2NHCH_2$ (ring)	Trimethylene imine	FIR, R	Double minimum potential function[a]. Barrier from lowest minimum = 441 cm^{-1}. Energy difference between minima = 91 cm^{-1}. N-*d* species studied	89, 90)
$C_4H_6O_2$	$CHCHCH_2OCH_2$ (ring), CH–O–CH epoxide bridge	3,6-Dioxabicyclo[3.1.0]hexane	FIR, R, MW	Single minimum potential function[a]. Minimum corresponds to a boat conformation	92, 93, 96)
$C_4H_6O_2$	CH_2COCH (ring), C=O, CH–CH_3	β-Butyrolactone	MW, FIR	Essentially planar. No potential function determined	152, 153)

Table 4.18C. (continued)

Formula	Structure	Name	Techniques	Comments	Ref.
C_4H_7Br	$CH_2CH_2CH_2CH$ (ring), Br on CH	Bromocyclobutane	MW, FIR, R	Single minimum potential function. Puckered ring with bromine equatorial	154–157)
C_4H_7Cl	$CH_2CH_2CH_2CH$ (ring), Cl on CH	Chlorocyclobutane	MW, FIR, R	Same as bromocyclobutane	157–160)
C_4H_7F	$CH_2CH_2CH_2CH$ (ring), F on CH	Fluorocyclobutane	MW, FIR, R	Same as bromocyclobutane	157–160)
C_4H_7N	$CH{=}CHCH_2NHCH_2$ (ring)	2,5-Dihydropyrrole	FIR, R, MW	Double minimum potential function with a barrier of the order of the zero point energy. N-*d* species studied	89, 90, 161)
C_4H_9N	$CH_2CH_2CCH_2$ (ring), NH_2 on C	Cyclobutyl amine	FIR, R	Single minimum potential function determined. ND_2 species studied	162)
C_5H_7N	$CH_2CH_2CHCH_2$ (ring), CN on CH	Cyanocyclobutane	FIR, MW	Single-minimum potential function determined. Puckered ring, CN equatorial	157,163, 164)
C_5H_8O	$CHCHCH_2CH_2CH_2$ (ring), O bridging CHCH	6-Oxabicyclo[3.1.0]hexane	FIR, R, MW	Same as 3,6-dioxabicyclo[3.1.0]-hexane	89, 91, 92, 96)
C_5H_8O	$CHCHCH_2OCH_2$ (ring), CH_2 bridging CHCH	3-Oxabicyclo[3.1.0]hexane	FIR, R, MW	Same as 3,6-dioxabicyclo[3.1.0]-hexane	92, 94, 96)

C_5H_8S	$CHCHCH_2CH_2CH_2$ \/ S	6-Thiabicyclo[3.1.0]hexane	MW	Boat conformation	165)
C_6H_8O	$CHCHCH_2CH{=}CHCH_2$ \/ O	7-Oxabicyclo[4.1.0]hept-3-ene	FIR, MW	Slightly anharmonic potential function determined. Boat conformation	166, 167)
C_6H_8O	$CH{-}CHOCH_2CH_2$ | | $CH{=}CH$	2-Oxabicyclo[3.2.0]hept-6-ene	FIR, R	Single-minimum potential function	168)
C_6H_{10}	$CHCHCH_2CH_2CH_2$ \/ CH_2	Bicyclo[3.1.0]hexane	FIR, R, MW	Same as 3,6-dioxabicyclo[3.1.0]-hexane	92, 95, 96)
C_7H_{10}	$CHCHCH_2CH_2CH_2$ | \\ $CH{=}CH$	Bicyclo[3.2.0]hept-6-ene	FIR, R	Single-minimum potential function	168)

[a] The potential function for trimethylene imine is unequivocally determined to have two distinct minima by the data. Similarly, the potential function for 3,6-dioxabicyclo[3.1.0]hexane has only a single minimum. Several molecules, e.g., fluorocyclobutane, have potential functions which exhibit a shallow ($<< 100$ cm^{-1}) second minimum at a position well above where the data end. Since there is no direct evidence for this second minimum and it is determined by a substantial extrapolation of the data, we have designated these cases "single-minimum" although this may not be precisely the case.

Table 4.18D. Five-membered and pseudo-five-membered ring molecules

Formula	Structure	Name	Techniques	Comments	Ref.
$C_2H_3FO_3$	$\overline{OCH_2OCHO}$ (ring), F on CH	Fluoroethylene ozonide	MW	Half-chair ring, fluorine axial	169)
$C_2H_4O_3$	$\overline{OCH_2OCH_2O}$ (ring)	Ethylene ozonide	MW	Half-chair ring conformation. Several isotopic species studied	170–172)
$C_3H_4O_3$	$\overline{CH_2OCOCH_2}$ (ring), =O on C	Ethylene carbonate	MW	Half-chair ring conformation. Splitting of MW transitions characteristic of a double minimum potential function was observed	173)
$C_3H_6O_2$	$\overline{CH_2OCH_2OCH_2}$ (ring)	1,3-Dioxolane	FIR, MW	One dimensional periodic function determined. Pseudorotational barrier = 45 cm^{-1}. C_s conformer 10 cm^{-1} more stable than C_2 conformer	36, 37, 100) 174)
$C_3H_6O_3$	$\overline{OCH_2OCHO}$ (ring), CH_3 on CH	Propylene ozonide	MW	Half-chair ring conformation with methyl equatorial. Isotopic species studied	175)
$C_4H_4O_2$	$\overline{OCH{=}CHOCH{=}CH}$ (ring)	1,4-Dioxadiene	FIR	Potential surface in bending and twisting coordiantes determined. (See also Table 4.11)	79, 84)

C_4H_6O	$CH{=}CHCH_2OCH_2$ (ring)	2,5-Dihydrofuran	FIR, MW, MIR, R	Potential surface in bending and twisting coordinates determined. (See also Table 4.11)	10, 12, 30, 78) 79, 80)
$C_4H_6O_2$	$CH_2CH_2C({=}O)OCH_2$ (ring)	γ-Butyrolactone	MW	Non-planar	176)
$C_4H_6O_2$	$CH{=}CHOCH_2CH_2O$ (ring)	1,4-Dioxene	FIR, R, MW	Two dimensional potential surface determined. The two equivalent half chair (C_2) forms interconvert via the half boat (C_s) forms	104–108)
C_4H_8O	$CH_2CH_2OCH_2CH_2$ (ring)	Tetrahydrofuran	FIR, MW	Periodic pseudo-rotational potential function determined. See 1,3-dioxolane ($C_3H_6O_2$)	6, 36–38) 100)
$C_4H_8O_3$	$OCH(CH_3)OCH(CH_3)O$ (ring)	Butylene ozonide	MW	Half-chair ring with methyls equatorial. Isotopic species studied	175)
C_4H_8S	$CH_2CH_2SCH_2CH_2$ (ring)	Tetrahydrothiophene	MW	Half chair ring conformation	177)
C_4H_8Se	$CH_2CH_2SeCH_2CH_2$ (ring)	Selenacyclopentane	MW	Half chair ring conformation	178)
$C_4H_{10}Ge$	$CH_2CH_2GeH_2CH_2CH_2$ (ring)	Germacyciopentane	FIR, MIR, R, MW	Two-dimensional potential function determined. The two half chair (C_2) conformers interconvert via the planar conformation. Barrier = 1454 cm^{-1}. (See cyclopentanone C_5H_8O)	179–181)

Table 4.18D. (continued)

Formula	Structure	Name	Techniques	Comments	Ref.
$C_4H_{10}Si$	$\overline{CH_2CH_2SiH_2CH_2CH_2}$	Silacyclopentane	FIR, R, MW, MIR	Same as germacyclopentane. Barrier = 1414 cm^{-1}. (See cyclopentanone C_5H_8O)	182–185)
C_5H_8	$\overline{CH{=}CHCH_2CH_2CH_2}$	Cyclopentene	FIR, MIR, R, MR, MW	See 2,5-dihydrofuran (C_4H_6O)	14, 20, 21, 48, 49, 78–82)
C_5H_8O	$\overline{CH_2CH_2C({=}O)CH_2CH_2}$	Cyclopentanone	FIR, MW	Two dimensional potential energy surface determined. (See germacyclopentane) Several deuterated species studied	10, 102, 103)
C_5H_8O	$\overline{CH{=}CHCH_2OCH_2CH_2}$	3,4-Dihydropyran	MW	Half chair conformation	186)
C_5H_8O	$\overline{CH{=}CHOCH_2CH_2CH_2}$	2,3-Dihydropyran	FIR, MW	(See 1,4-dioxene $C_4H_6O_2$)	104)
C_5H_9N	$\overline{CH{=}CHCH_2NHCH_2CH_2}$	1,2,3,6-Tetrahydropyridine	MW	Mixture of half-chair-N-H equatorial and half-chair N-H axial. The equatorial conformer is slightly more abundant. N-*d* species studied	187)
C_5H_{10}	$\overline{CH_2CH_2CH_2CH_2CH_2}$	Cyclopentane	MIR, R	Essentially pure pseudorotation. Barrier to planarity = 1824 cm^{-1}. Isotopic species studied	7, 97, 188, 189)
C_6H_{10}	$\overline{CH{=}CHCH_2CH_2CH_2CH_2}$	Cyclohexene	MW, R	See 1,4-dioxene ($C_4H_6O_2$)	105, 190, 191)
C_6H_{10}	$\overline{CH_2CH_2C({=}CH_2)CH_2CH_2}$	Methylenecyclopentane	MW	Half chair conformation	192)

VI References

1. Bell, R. P.: Proc. Roy. Soc. (London) *A 183*, 328 (1945)
2. Rathjens, G. W., Freeman, N. R., Gwinn, W. D., Pitzer, K. S.: J. Am. Chem. Soc. *75*, 5634 (1953)
3. Kilpatrick, J. E., Pitzer, K. S., Spitzer, R.: J. Am. Chem. Soc. *69*, 2483 (1947)
4. Danti, Alfred: Studies in Far Infrared spectroscopy, Ph. D. Thesis, Massachusetts Institute of Technology, September, 1958
5. Danti, A., Lafferty, W. J., Lord, R. C.: J. Chem. Phys. *33*, 294 (1960)
6. Lafferty, W. J., Robinson, D. W., St. Louis, R. V., Russell, J. L., Strauss, H. L.: J. Chem. Phys. *42*, 2915 (1965)
7. Durig, J. R., Wertz, D. W.: J. Chem. Phys. *49*, 2118 (1968)
8. For example, Hard, T. M., Lord, R. C.: Appl. Optics *7*, 589 (1968)
9. For example, Jokisaari, J., Kauppinen, J.: J. Chem. Phys. *59*, 2260 (1973)
10. Carreira, L. A., Lord, R. C.: J. Chem. Phys. *51*, 3225 (1969)
11. Moeller, K. D., Rothschild, W. G.: Far-infrared spectroscopy. Wiley-Interscience: New York: 1971
12. Ueda, T., Shimanouchi, T.: J. Chem. Phys. *47*, 4042 (1967)
13. Miller, F. A., Capwell, R. J.: Spectrochim. Acta *27A*, 947 (1971)
14. Chao, T. H., Laane, J.: Chem. Phys. Lett. *14*, 595 (1972)
15. Gibson, J. S., Harris, D. O.: J. Chem. Phys. *57*, 2318 (1972)
16. Scharpen, L. H., Laurie, V. W.: J. Chem. Phys. *49*, 221 (1968)
17. Scharpen, L. H., Laurie, V. W.: J. Chem. Phys. *49*, 3041 (1968)
18. Harris, D. O., Harrington, H. W., Luntz, A. C., Gwinn, W. D.: J. Chem. Phys. *44*, 3467 (1966)
19. Gwinn, W. D., Gaylord, A. S.: Spectroscopic Studies of Ring-Puckering Motions, in MTP International Reviews of Science, Physical Chemistry Series 2, Vol. 3 Spectroscopy, Ramsay, D. A., (ed.) London and Baltimore: Butterworths and University Park Press 1976
20. Scharpen, L. H.: J. Chem. Phys. *48*, 3552 (1968)
21. Pickett, H. L.: J. Chem. Phys. *56*, 1715 (1972)
22. Meyer, R., Günthard, H. H.: J. Chem. Phys. *49*, 1510 (1968)
23. Kemble, E. C.: The fundamental principles of quantum mechanics. New York: McGraw-Hill Book Co. 1937
24. Laane, J.: Appl. Spectros. *24*, 73 (1970)
25. Wilson, E. B., Decius, J. C., Cross, P. C.: Molecular vibrations. New York: McGraw-Hill Book Co. 1955
26. Wicke, B. G., Harris, D. O.: J. Chem. Phys. *64*, 5236 (1976)
27. Malloy, T. B., Jr., Lafferty, W. J.: J. Mol. Spectroscopy *54*, 20 (1975)
28. Reid, C.: J. Mol. Spectroscopy *36*, 183 (1970)
29. Chan, S. I., Stelman, D.: J. Mol. Spectroscopy *10*, 278 (1963)
30. Carreira, L. A., Mills, I. M., Person, W. B: J. Chem. Phys. *56*, 1444 (1972)
31. For example, Ralston, A.: A first course in numerical analysis. New York: McGraw-Hill Book Co. 1965
32. Chan, S., Zinn, J., Gwinn, W. D.: J. Chem. Phys. *34*, 1319 (1961)
33. Stone, J. M. R., Mills, I. M.: Mol. Phys. *18*, 631 (1970)
34. Harris, D. O., Engerholm, G. G., Tolman, C. A., Luntz, A. C., Keller, R. A., Kim, H., Gwinn, W. D.: J. Chem. Phys. *50*, 2438 (1969)
35. Ikeda, T., Lord, R. C., Malloy, T. B., Ueda, T.: J. Chem. Phys. *56*, 1434 (1972)
36. Greenhouse, J. A., Strauss, H. L.: J. Chem. Phys. *50*, 124 (1969)
37. Davidson, R., Warsop, P. A.: J. Chem. Soc., Farad. Trans. II *68*, 1875 (1972)
38. Engerholm, G. G., Luntz, A. C., Gwinn, W. D., Harris, D. O.: J. Chem. Phys. *50*, 2446 (1969)
39. Lewis, J. D., Chao, T. H., Malloy, T. B., Laane, J.: J. Mol. Structure *12*, 427 (1972)
40. Laane, J.: Pseudorotation of five-membered rings, Chap. 2, in: Vibrational spectra and structure, Vol. 1. Durig, J. R. (ed.). New York: Marcel Dekker, Inc. 1972, pp. 25–50

41. Blackwell, C. S., Lord, R. C.: Far-infrared spectra of four-membered-ring compounds, Chap. 1, in: Vibrational spectra and structure, Vol. 1, Durig, J. R. (ed.). New York: Marcel Dekker, Inc. 1972, pp. 1–24
42. Wurrey, C. J., Durig, J. R., Carreira, L. A.: Gas-phase raman spectroscopy of anharmonic vibrations. Chap. 4, in: Vibrational spectra and structure, Vol. 5. Durig, J. R. (ed.). Amsterdam-Oxford-New York: Elsevier Sci. Pub. Co. 1976, pp. 121–277
43. Blackwell, C. S., Lord, R. C.: J. Mol. Spectroscopy *55*, 460 (1975)
44. Avirah, T. K., Cook, R. L., Malloy, T. B.: J. Mol. Spectroscopy *55*, 464 (1975)
45. Chan, S. I., Zinn, J., Fernandez, J., Gwinn, W. D.: J. Chem. Phys. *33*, 1643 (1960)
46. Chan, S. I., Zinn, J., Gwinn, W. D.: J. Chem. Phys. *34*, 1319 (1961)
47. Chan, S. I., Borgers, T. R., Russell, J. W., Strauss, H. L., Gwinn, W. D.: J. Chem. Phys. *44*, 1103 (1966)
48. Ueda, T., Shimanouchi, T.: J. Chem. Phys. *47*, 5018 (1967)
49. Green, W. H.: J. Chem. Phys. *52*, 2156 (1970)
50. Wieser, H., Danyluk, M., Kydd, R. A.: J. Mol. Spectroscopy *43*, 382 (1972)
51. Kiefer, W., Bernstien, H. J., Wieser, H., Danyluk, M.: J. Mol. Spectroscopy *43*, 393 (1972)
52. Kydd, R. A., Wieser, H., Danyluk, M.: J. Mol. Spectroscopy *44*, 14 (1972)
53. Wieser, H., Danyluk, M.: Can. J. Chem. *50*, 2761 (1972)
54. Wieser, H., Danyluk, M., Kiefer, W., Bernstein, H. J.: Can. J. Chem. *50*, 2771 (1972)
55. Kiefer, W., Bernstein, H. J., Danyluk, M., Wieser, H.: Chem. Phys. Letters *12*, 605 (1972)
56. Wieser, H., Danyluk, M., Kydd, R. A., Kiefer, W., Bernstein, H. J.: J. Chem. Phys. *61*, 4380 (1974)
57. Creswell, R. A., Mills, I. M.: J. Mol. Spectroscopy *52*, 392 (1974)
58. Mallinson, P. D., Robiette, A. G.: J. Mol. Spectroscopy *52*, 413 (1974)
59. Creswell, R. A.: Mol. Phys. *30*, 217 (1975)
60. Kivelson, D., Wilson, E. B., Jr.: J. Chem. Phys. *20*, 1575 (1952)
61. Watson, J. K. G.: J. Chem. Phys. *46*, 1935 (1967)
62. Watson, J. K. G.: J. Chem. Phys. *48*, 181 (1968)
63. Borgers, T. R., Strauss, H. L.: J. Chem. Phys. *45*, 947 (1966)
64. Durig, J. R., Lord, R. C.: J. Chem. Phys. *45*, 61 (1966)
65. Durig, J. R., Shing, A. C., Carreira, L. A., Li, Y. S.: J. Chem. Phys. *57*, 4398 (1972)
66. Butcher, S. S., Costain, C. C.: J. Mol. Spectroscopy *15*, 40 (1965)
67. Wieser, H., Duckett, J. A.: J. Mol. Spectroscopy *50*, 443 (1974)
68. Wieser, H., Duckett, J. A., Kydd, R. A.: J. Mol. Spectroscopy *51*, 115 (1974)
69. Wieser, H., Kydd, R. A.: J. Raman Spectrosc. *4*, 401 (1976)
70. Malloy, T. B., Fisher, F., Hedges, R. M.: J. Chem. Phys. *52*, 5325 (1970)
71. Laane, J., Lord, R. C.: J. Chem. Phys. *48*, 1508 (1968)
72. Lewis, J. D., Chao, T. H., Laane, J.: J. Chem. Phys. *62*, 1932 (1975)
73. Blanke, J. F., Chao, T. H., Laane, J.: J. Mol. Spectroscopy *38*, 483 (1971)
74. Pringle, W. C.: J. Chem. Phys. *54*, 4979 (1971)
75. Pettit, M. G., Gibson, J. S., Harris, D. O.: J. Chem. Phys. *53*, 3408 (1970)
76. Harvey, A. B., Durig, J. R., Morrisey, A. C.: J. Chem. Phys. *50*, 4949 (1969)
77. Ueda, T., Shimanouchi, T.: J. Chem. Phys. *49*, 470 (1968)
78. Malloy, T. B.: J. Mol. Spectroscopy *44*, 504 (1972)
79. Malloy, T. B., Carreira, L. A.: J. Chem. Phys. *71*, in press (1979)
79a. Laane, J., Lord, R. C.: J. Chem. Phys. *47*, 4941 (1967)
80. Durig, J. R., Carreira, L. A.: J. Chem. Phys. *56*, 4966 (1972)
81. Villareal, J. R., Bauman, L. E., Laane, J., Harris, W. C., Bush, S. F.: J. Chem. Phys. *63*, 3727 (1975)
82. Villareal, J. R., Bauman, L. E., Laane, J.: J. Phys. Chem. *80*, 1172 (1976)
83. Laane, J.: private communication
84. Lord, R. C., Rounds, T. C.: J. Chem. Phys. *58*, 4344 (1973)
85. Butcher, S. S.: J. Chem. Phys. *42*, 1830 (1965)
86. Carreira, L. A., Carter, R. O., Durig, J. R.: J. Chem. Phys. *59*, 813 (1973)
87. Carreira, L. A.: unpublished results

88. Laane, J., Lord, R. C.: J. Mol. Spectroscopy *39*, 340 (1971)
89. Carreira, L. A., Lord, R. C.: J. Chem. Phys. *51*, 2735 (1969)
90. Carreira, L. A., Carter, R. O., Durig, J. R.: J. Chem. Phys. *57*, 3384 (1972)
91. Lafferty, W. J.: J. Mol. Spectroscopy *36*, 84 (1970)
92. Lord, R. C., Malloy, T. B., Jr.: J. Mol. Spectroscopy *46*, 358 (1973)
93. Creswell, R. A., Lafferty, W. J.: J. Mol. Spectroscopy *46*, 371 (1973)
94. Malloy, T. B., Jr.: J. Mol. Spectroscopy *49*, 432 (1974)
95. Cook, R. L., Malloy, T. B., Jr.: J. Am. Chem. Soc. *96*, 1703 (1974)
96. Lewis, J. D., Laane, J., Malloy, T. B., Jr.: J. Chem. Phys. *61*, 2342 (1974)
97. Carreira, L. A., Jiang, G. J., Person, W. B., Willis, J. N., Jr.: J. Chem. Phys. *56*, 1440 (1972)
98. Pitzer, K. S., Donath, W.: J. Am. Chem. Soc. *81*, 3213 (1959)
99. Durig, J. R., Wertz, D. W.: J. Chem. Phys. *49*, 675 (1968)
100. Lord, R. C., Ueda, T.: unpublished results
101. Sont, W. N., Wieser, H.: Thirty Second Annual Symposium on Molecular Spectroscopy, Columbus, Ohio, 1977 (Paper FA 4)
102. Ikeda, T., Lord, R. C.: J. Chem. Phys. *56*, 4450 (1972)
103. Kim, H., Gwinn, W. D.: J. Chem. Phys. *51*, 1815 (1969)
104. Lord, R. C., Rounds, T. C., Ueda, T.: J. Chem. Phys. *57*, 2572 (1972)
105. Durig, J. R., Carter, R. O., Carreira, L. A.: J. Chem. Phys. *60*, 3098 (1974)
106. Larkin, R. H., Lord, R. C.: J. Am. Chem. Soc. *95*, 5129 (1973)
107. Larkin, R. H., Lord, R. C.: J. Am. Chem. Soc. *96*, 1643 (1974)
108. Wells, J. A., Malloy, T. B., Jr.: J. Chem. Phys. *60*, 2132 (1974)
109. Strauss, H. L., Pickett, H. M.: J. Chem. Phys. *53*, 376 (1970)
110. Strauss, H. L.: J. Chem. Educ. *48*, 221 (1971)
111. Strauss, H. L., Bocian, D. F., Pickett, H. M., Rounds, T. C.: J. Am. Chem. Soc. *97*, 687 (1975)
112. Strauss, H. L., Bocian, D. F.: J. Am. Chem. Soc. *99*, 2866, 2876 (1977)
113. Rounds, T. C., Strauss, H. L.: Vibrational spectroscopy of medium rings. Chap. 2, in: Vibrational spectra and structure, Vol. 7. Durig, J. R. (ed.). Amsterdam-Oxford-New York: Elsevier Sci. Pub. Co. 1978
114. See for example, McDowell, R. S., Galbraith, H. W., Cantrell, C. D., Nereson, N. G., Hinkley, E. D.: J. Mol. Spectros. *68*, 288 (1977)
115. See for example, Henesian, N. A., Byer, R. L.: High-resolution CARS line shape function. In: Proc. Tenth Int'l. Quantum Electronics Conf., Atlanta, GA May 29-June 1, 1978, Paper G9, p. 648
116. Durig, J. R., Lord, R. C.: Spectrochim. Acta *19*, 769 (1963)
117. Irwin, R. M., Cooke, J. M., Laane, J.: J. Amer. Chem. Soc. *99*, 3273 (1977)
118. McKown, G. L., Beaudet, R. A.: J. Chem. Phys. *55*, 3105 (1971)
119. Pringle, W. C., Jr., Meinzer, A. L.: J. Chem. Phys. *61*, 2071 (1974)
120. Durig, J. R., Morrissey, A. C., Harris, W. C.: J. Mol. Struct. *6*, 375 (1970)
121. Boone, D. W., Britt, C. O., Boggs, J. E.: J. Chem. Phys. *43*, 1190 (1965)
122. Durig, J. R., Harris, W. C.: Spectrochim. Acta *27A*, 947 (1971)
123. Durig, J. R., Li, Y. S., Hudgens, B. A., Cohen, E. A., Jr.: J. Mol. Spectrosc. *63*, 459 (1976)
124. Moning, F., Driezler, H., Rudolph, H. D.: Z. Naturforschung *229*, 1471 (1967)
125. Luntz, A.: J. Chem. Phys. *50*, 1109 (1969)
126. Stigliani, W. M., Laurie, V. W.: J. Mol. Spectrosc. *62*, 85 (1976)
127. Durig, J. R., Morrissey, A. C.: J. Chem. Phys. *45*, 1269 (1969)
128. Gibson, J. S., Harris, D. O.: J. Chem. Phys. *52*, 5234 (1970)
129. Duckett, J. A., Smithson, T. L., Wieser, H.: J. Mol. Spectrosc. *69*, 159 (1978)
130. Avirah, T. K., Cook, R. L., Malloy, T. B., Jr.: J. Mol. Spectrosc. *54*, 231 (1975)
131. Pringle, W. C., Jr., Meinzer, A. L.: J. Chem. Phys. *57*, 2920 (1972)
132. Carreira, L. A., Odom, J. O., Durig, J. R.: J. Chem. Phys. *59*, 4955 (1973)
133. Pringle, W. C., Appeloff, C., Jordan, K. W.: J. Mol. Spectrosc. *55*, 351 (1975)
134. Gaylord, A. S., Pringle, W. C., Jr.: J. Chem. Phys. *59*, 4674 (1973)
135. Dorris, K. L., Britt, C. O., Boggs, J. E.: J. Chem. Phys. *44*, 1352 (1966)

136. White, W. F., Boggs, J. E.: J. Chem. Phys. *54*, 4714 (1971)
137. Durig, J. R., Riethmiller, S., Li, Y. S.: J. Chem. Phys. *61*, 253 (1974)
138. Durig, J. R., Karriker, J. M., Harris, W. C.: J. Chem. Phys. *52*, 6096 (1970)
139. Durig, J. R., Carreira, L. A., Laane, J.: J. Mol. Struct. *21*, 281 (1974)
140. Green, W. H.: J. Chem. Phys. *50*, 1619 (1969)
141. Durig, J. R., Carter, R. O., Carreira, L. A.: J. Chem. Phys. *59*, 2249 (1973)
142. Durig, J. R., Li, Y. S., Tong, C. K.: J. Chem. Phys. *56*, 5692 (1972)
143. Green, W. H., Harvey, A. B.: J. Chem. Phys. *49*, 177 (1968)
144. Laane, J.: J. Chem. Phys. *50*, 776 (1969)
145. Laane, J.: J. Chem. Phys. *52*, 358 (1970)
146. Lewis, J. D., Laane, J.: J. Mol. Spectrosc. *53*, 417 (1974)
147. Bevan, J. W., Legon, A. C.: Chem. Comm. *1971*, 1136
148. Bevan, J. W., Legon, A. C.: J. Chem. Soc. Farad, Trans. II, *69*, 902 (1973); J. Chem. Soc. Farad. Trans. II, *69*, 916 (1973)
149. Chao, T. H., Laane, J.: J. Mol. Spectrosc. *48*, 366 (1973)
150. Chadwick, D., Legon, A. C., Millen, D. J.: Chem. Comm. *1969*, 1130
151. Bevan, J. W., Legon, A. C., Millen, D. J.: Chem. Comm. *1974*, 659
152. Durig, J. R., Morrissey, A. C.: J. Mol. Struct. *2*, 377 (1968)
153. Coffey, D., Hershberger, M. V.: J. Mol. Spectrosc. *59*, 28 (1976)
154. Rothschild, W. G., Dailey, B. P.: J. Chem. Phys. *36*, 2931 (1962)
155. Rothschild, W. G.: J. Chem. Phys. *44*, 2213 (1966)
156. Durig, J. R., Shing, A. C., Carreira, L. A.: J. Mol. Struct. *17*, 423 (1973)
157. Blackwell, C. S., Carreira, L. A., Durig, J. R., Karriker, J. M., Lord, R. C.: J. Chem. Phys. *56*, 1706 (1972)
158. Kim, H., Gwinn, W. D.: J. Chem. Phys. *44*, 865 (1966)
159. Rothschild, W. G.: J. Chem. Phys. *45*, 1214 (1966)
160. Durig, J. R., Carreira, L. A., Willis, J. N.: J. Chem. Phys. *57*, 2755 (1972)
161. Nave, C. R., Pullen, K. P.: Chem. Phys. Lett. *12*, 499 (1972)
162. Kalasinsky, V. F., Guirgis, G. A., Durig, J. R.: J. Mol. Struct. *39*, 51 (1977)
163. Fong, M. Y., Harmony, M. D.: J. Chem. Phys. *58*, 4260 (1973)
164. Durig, J. R., Carreira, L. A., Lafferty, W. J.: J. Mol. Spectrosc. *46*, 187 (1973)
165. Mjöberg, P. S., Ralowski, W. M., Ljunggren, S. O., Bäckvall, J. E.: J. Mol. Spectrosc. *60*, 179 (1976)
166. Malloy, T. B., Jr.: J. Chem. Phys. *65*, 2538 (1976)
167. Chao, S., Cook, R. L., Malloy, T. B. Jr.: J. Chem. Phys. *68*, 4027 (1978)
168. Villareal, J. R., Laane, J.: J. Chem. Phys. *68*, 3298 (1978)
169. Lattimer, R. P., Mazur, U., Kuczkowski, R. L.: J. Amer. Chem. Soc. *98*, 4012 (1976)
170. Gillies, C. W., Lattimer, R. P., Kuczkowski, R. L.: J. Amer. Chem. Soc. *96*, 1536 (1974)
171. Gillies, C. W., Kuczkowski, R. L.: J. Amer. Chem. Soc. *94*, 6336 (1972)
172. Kuczkowski, R. L., Gillies, C. W., Gallaher, K. L.: J. Mol. Spectrosc. *60*, 361 (1976)
173. Wang, I., Britt, C. O., Boggs, J. E.: J. Amer. Chem. Soc. *87*, 4950 (1965)
174. Baron, P. A., Harris, D. O.: J. Mol. Spectrosc. *49*, 70 (1974)
175. Lattimer, R. P., Kuczkowski, R. L., Gillies, C. W.: J. Amer. Chem. Soc. *96*, 348 (1974)
176. Durig, J. R., Li, Y. S., Tong, C. C.: J. Mol. Struct. *18*, 269 (1973)
177. Mainleev, A. K., Pozdeev, N. M.: Zh. Strukt. Khim. *10*, 747 (1969)
178. Mainleev, A. K., Magdeseria, N. N., Pozdeev, N. M.: Zh. Strukt. Khim. *11*, 1124 (1970)
179. Durig, J. R., Willis, J. N.: J. Chem. Phys. *52*, 6108 (1970)
180. Thomas, E. C., Laurie, V. W.: J. Chem. Phys. *51*, 4327 (1969)
181. Durig, J. R., Li, Y. S., Carreira, L. A.: J. Chem. Phys. *58*, 2393 (1973)
182. Laane, J.: J. Chem. Phys. *50*, 1946 (1969)
183. Durig, J. R., Willis, J. N.: J. Mol. Spectrosc. *32*, 320 (1969)
184. Durig, J. R., Natter, W. J., Kalasinsky, V. F.: J. Chem. Phys. *67*, 4756 (1977)
185. Durig, J. R., Lafferty, W. J., Kalasinsky, V. F.: J. Phys. Chem. *80*, 1199 (1976)
186. Wells, J. A., Malloy, T. B., Jr.: J. Chem. Phys. *60*, 3987 (1974)
187. Chao, S., Avirah, T. K., Cook, R. L., Malloy, T. B., Jr: J. Phys. Chem. *80*, 1141 (1976)

188. Bauman, L. E., Chao, T. H., Laane, J.: Thirty First Annual Symposium on Molecular Spectroscopy, The Ohio State University, Columbus, Ohio. June, 1976 (Paper FC3)
189. Chao, T. H., Laane, J.: J. Mol. Spectrosc. *70*, 357 (1978)
190. Scharpen, L. H., Wollrab, J. E., Ames, D. P.: J. Chem. Phys. *49*, 2368 (1968)
191. Ogata, T., Kozima, K.: Bull. Chem. Soc. Japan *42*, 1263 (1969)
192. Durig, J. R., Li, Y. S., Carreira, L. A.: J. Chem. Phys. *57*, 1896 (1972)

Received August 4, 1978

A New Approach to the Hamiltonian of Nonrigid Molecules

Georg Ole Sørensen

Chemical Laboratory V, University of Copenhagen, H. C. Ørsted Institute, 5, Universitetsparken, DK-2100 Copenhagen Ø, Denmark

Table of Contents

1 Introduction

The dynamics of nonrigid molecules has been studied with increasing interest in recent years. This is a natural consequence of the increasing amount of very precise data for these molecules made available by the developments within high resolution spectroscopy. Such data require detailed analysis. Interest has also been stimulated by the fact that developments within computer technology have allowed an attempt at solving the involved numerical problems.

Theoretical formulations have now reached a level which allows the possibility of standardizing the treatment of nonrigid molecules in a way which is very similar to the treatment applied to rigid molecules. This is where the present paper may hopefully contribute.

In rigid molecules the vibrational amplitudes are so small that the vibration-rotation spectra can be analyzed in great detail within the formalism of Wilson and Howard[1-3] using the fully elaborated perturbation scheme of Amat and Nielsen[4]. A review of the method has recently been given by Mills[5]. The latest development concerns the anharmonic force constants which, according to Hoy, Mills and Strey[6], should be defined as the partial derivatives of the potential function with respect to structural parameters, i.e. curvilinear coordinates as opposed to the rectilinear coordinates used in the Hamiltonian. This is made possible employing the nonlinear transformation between the two classes of coordinates which can be evaluated by the methods presented by Hoy et al.[6]. This development is equally important when small amplitude motions in nonrigid molecules are considered. An alternative treatment has been suggested by Quade[7] based on a rigorous use of curvilinear coordinates in formulating the Hamiltonian. Comments on this method will be presented in a later section.

In nonrigid molecules one or more internal motions take place with distortions so large that the ordinary treatment is inadequate or even breaks down. Thus, in the case of large amplitude vibrations, bending, inversion or ring puckering modes in particular, a Hamiltonian based on rectilinear coordinates is still exact, but the perturbation treatment converges only very slowly because of the change in order of magnitude of certain terms. When internal rotation is considered, it is no longer appropriate to describe the internal motions as displacements of atoms relative to a single unique equilibrium configuration. Therefore both types of nonrigidity require special treatments.

Such treatments have generally been based on semirigid models, where the small amplitude vibrations are neglected so that only the overall rotation and the relative motion of a few (usually two) rigid groups in the molecule are considered. The methods applied to internal rotation have been described in detail by Dreizler[8, 9], whereas a complete review of the many different approaches to the large amplitude vibration problem is very difficult to present. The development of the theory may, however, be followed in a few illustrative papers. In the pioneering papers by Thorson and Nakagawa[10], Dixon[11] and Johns[12] and also in later works[13, 14] an isolated bending was studied using an approximated kinetic energy expression with constant reduced mass, equivalent of using rectilinear coordinates. Various types of anharmonic potential functions, including double minimum types, were studied. Within this approach to large amplitude bendings it has been possible to quite accurately account for vibration-rotation coupling effects as demonstrated in the recent papers by Duckett, Robiette and Mills[15].

Aiming at a more detailed study of large amplitude potential, it is necessary to use curvilinear coordinates and hence to use a coordinate-dependent reduced mass. Large amplitude bending in quasilinear molecules was discussed by Shinkle and Coon[16]; including overall rotation by Hougen, Bunker and Johns[17]. Using the principles outlined in the latter paper, several types of large amplitude internal motions have been treated: Bendings in HCN and H_2O (Bunker and Stone[18], Bunker and Landsberg[19]), in HCNO (Stone[20], Bunker et al.[21]) and in C_3O_2 (Weber and Ford[22]), inversions in H_2CO (Moule and Rao[23]), in NH_3 and deuterated NH_3 (Papoušek et al.[24], Daniels et al.[25]) and in CH_3NH_2(Kreglewski[26]). More special internal motions have been discussed by Henderson and Ewing[27] in relation to Van der Waal molecules and by Istomin[28] concerning a migration of the lithium atom around the cyanide group in LiCN. Also a more elaborated model for ring puckering has appeared (Malloy et al.[29, 30]) which was applied to cyclopentene, by Villarreal et al.[31].

For the treatment of the internal rotation of a non-axially symmetric top an angle-dependent reduced moment of inertia must be introduced. In this way nitroethylene was studied by Bauder et al.[32] and butadiene by Carreira[33]. An angle-dependent reduced moment of inertia has also been introduced in the study of methyl group internal rotation to account for structural relaxation[34, 35].

When considering the results of such treatments it should be borne in mind that there is a principal equivalence between applying the semirigid rotor model to non-rigid molecules and applying the rigid rotor model to rigid molecules. In both cases we must realize that the parameters of the model Hamiltonian are effective constants for the particular state of the neglected small amplitude vibrations. For the rigid molecules a rigorous method of calculating all the effective constants is provided by the general formalism mentioned above, but this is not so for the nonrigid molecules. The question has been discussed, however, in relation to specific problems. Thus perturbation effects are incorporated in the internal rotation treatment by Kirtman[36], in the treatment of H_2O by Hoy and Bunker[37] and in the analysis of centrifugal distortion and Coriolis coupling in NH_3 by Spirko et al.[38]. In the present paper general expressions of the effective constants will be given which are applicable to molecules with large amplitude internal motions of any type.

Hamiltonians will only be discussed within the framework of the Born-Oppenheimer approximation. The fundamental problem is then to make a convenient choice of generalized coordinates and momenta which ensures rapid convergence in the expansion of the Hamiltonian. What remains after this is the specific work of deriving which involves algebraic problems only. However, this work is so difficult and tedious that any formalism which helps to simplify this part may be important for future progress. For this reason Sect. 2 is devoted to the discussion of such a formalism.

The basic principle of this is to derive the Hamiltonian form of the kinetic energy directly from the transformation of the momenta. Hence, with linear Cartesian momenta, P_i; $i = 1, 2, \ldots 3N$, and generalized (including quasi-generalized) momenta, π_j; $j = 1, 2, \ldots 3N$, we shall set up the square matrix $\{s_{ji}\}$ of the transformation,

$$P_i = \sum_j \pi_j s_{ji} \tag{1.1}$$

From this we obtain the transformed kinetic energy expression by substituting,

$$2T = \sum_i m_i^{-1} P_i^2 = \sum_{jj'} G_{jj'} \pi_j \pi_{jj'} \tag{1.2}$$

introducing the general G-matrix by

$$G_{jj'} = \sum_i m_i^{-1} s_{ji} s_{j'i} \tag{1.3}$$

This applies to the quantum mechanical operators as well.

In Sect. 3 the Wilson-Howard operator is discussed as an example of application. From this it appears that the Eckart conditions[39] can be inferred from arguments which are easily extended to Sayvetz conditions[40] of any type. The general derivation of Hamiltonians of nonrigid molecules can then be presented in Sect. 4, and an effective semirigid rotor Hamiltonian is formed by a Van Vleck transformation. Finally Sect. 5 gives a complete example of a calculation on a specific molecule, C_3.

2 Derivation of Hamiltonians

The methods used in setting up the complete vibration-large amplitude motion-rotation Hamiltonian are illustrated by several examples in the literature. Thus the case of methyl group internal rotation was discussed by Kirtman[36], and his method was applied by Iijima and Tsuchiya[41] to acetaldehyde and by Fleming and Banwell[42] to molecules with free or slightly hindered internal rotation. The treatment was extended by Quade[43] to cover an asymmetric internal rotor as well. The effects of a large amplitude bending or inversion have been extensively discussed[44–46], and particularly the work of Hougen, Bunker and Johns[17] on triatomic molecules has formed the basis for several attempts to treat more complicated systems[23–26, 47, 48].

In all these cases the Hamiltonian form of the kinetic energy was derived by procedures that are fundamentally similar to the original method for ordinary rigid molecules used by Wilson and Howard[1, 2]. A presentation of their treatment is now found as an essential part of most textbooks on vibration-rotation spectroscopy[49–52] and their notation is therefore assumed to be a widely accepted standard. For convenience we adopt a similar notation here, and in particular we shall use vectors of our ordinary three-dimensional space when discussing atomic positions, velocities and momenta.

The general features of the usual treatment will be summarized below, but first we will discuss properties of the generalized coordinates which are fundamental for any approach to the vibration-rotation Hamiltonian.

2.1 Generalized Coordinates

By the Born-Oppenheimer adiabatic approximation we obtain a molecular model in which the potential energy depends on structural variables of the nuclear framework only, whereas it is independent of the position of the molecule in space. Correspondingly it is convenient to use generalized coordinates which are divided into two classes, the internal coordinates determining the relative positions of the N atoms,

$$q_k;\ k = 1, 2, \ldots 3N - 6 \tag{2.1 a}$$

and six coordinates describing the rigid motions. It is necessary to distinguish between translational and rotational coordinates and here we shall take

$$R_X, R_Y, R_Z, \varphi, \theta, \chi \tag{2.1 b}$$

three position vector components for the center of mass and three Eulerian angles. They specify the position of a translating and rotating molecular reference coordinate system relative to a laboratory fixed system.

For a more explicit definition we introduce the following vectors of three-dimensional space,

$\mathbf{e}_F$; $F = X, Y, Z$, orthogonal unit vectors forming the basis of the Laboratory System, LS.
$\mathbf{e}_g$; $g = x, y, z$, orthogonal unit vectors forming the basis of the Molecular System, MS. (2.2)
$\mathbf{R}_\alpha$; $\alpha = 1, 2, \ldots N$, position vectors of the atoms from the origin of LS.
$\mathbf{r}_\alpha$, position vector of atom from the origin of MS.
$\mathbf{R}$, position vector of the center of mass from the origin of LS.

Vector components are indicated by adding subscripts $F = X, Y, Z$ or $g = x, y, z$, the labels of the LS- or MS-axes respectively, e.g.

$$R_X = \mathbf{e}_X \cdot \mathbf{R}, \quad r_{\alpha z} = \mathbf{e}_z \cdot \mathbf{r}_\alpha$$

The origin of the molecular system is chosen as the instantaneous center of mass. From the vector definitions above it therefore follows that

$$\mathbf{R}_\alpha = \mathbf{R} + \mathbf{r}_\alpha \tag{2.3}$$

and

$$\sum_\alpha m_\alpha \mathbf{r}_\alpha = \mathbf{0}, \quad \text{or} \quad \sum_\alpha m_\alpha r_{\alpha g} = 0, \quad g = x, y, z \tag{2.4}$$

The orientation of molecular axes are given by the directional cosines, Φ_{Fg},

$$\mathbf{e}_g = \sum_F \mathbf{e}_F\, \Phi_{Fg} \tag{2.5}$$

which are trigonometric functions of the Eulerian angles, φ, θ and χ (see App. I of Ref. 2). The positions of the atoms relative to the MS-system, as given by the components, $r_{\alpha g}$, are functions of the internal coordinates only.

These definitions allow us to express the relations between the generalized coordinates and the $3N$ Cartesian coordinates of an arbitrary configuration in the following formal way

$$R_{\alpha F} = R_F + \sum_g \Phi_{Fg}(\varphi, \theta, \chi) \cdot r_{\alpha g}(q_1, q_2, \ldots q_{3N-6})$$
$$\alpha = 1, 2, \ldots N, \qquad F = X, Y, Z, \qquad g = x, y, z \tag{2.6}$$

The dependence of $r_{\alpha g}$ on the internal coordinates is not restricted by requirements other than the center of mass conditions (2.4) and that Eq. (2.6) is invertible. In expressing the $r_{\alpha g}$ functions we may therefore also consider how the final Hamiltonian is influenced, so that we obtain an operator of optimum suitability characterized by e.g. rapid convergence of the perturbing terms. In this respect there are two particular concerns, the vibration-rotation interaction and the potential energy expansion.

The vibration-rotation interaction is the effect arising from coupling terms between angular and vibrational momenta as well as from the dependence of the rotational G-matrix elements (the μ-tensor) on the internal coordinates. The importance of this effect may to some extent be reduced provided an appropriate axis convention is used. The axis convention is the set of rules defining the orientation of the molecular axes, $\mathbf{e}_g$, $g = x, y, z$, relative to an arbitrary configuration as given by the position vectors, $\mathbf{R}_\alpha$, $\alpha = 1, 2, \ldots N$. These rules can be expressed in three relations between the $r_{\alpha g}$ components, similar to the center of mass conditions (2.4). We shall refer to these relations as "the axial constraints". Usually Eckart-conditions[39)] are imposed, but other possibilities may be considered.

Rapid convergence of the potential energy expansion depends on the type of internal coordinates involved. Of particular importance here is whether we use curvilinear coordinates that are close to the true geometrical variables (e.g. the valence coordinates comprising bond lengths and angles, etc.) or whether we use rectilinear coordinates. We shall here define rectilinear coordinates as a subset, $q_1, q_2, \ldots q_n$; $n \leqslant 3N-6$, of internal coordinates which enter only linearly in the expression for the $r_{\alpha g}$ components

$$r_{\alpha g} = a_{\alpha g} + \sum_{k=1}^{n} \frac{\partial r_{\alpha g}}{\partial q_k} q_k \tag{2.7}$$

It is assumed that $a_{\alpha g}$, as well as the partial derivatives, are functions of the remaining curvilinear coordinates, $q_{n+1}, q_{n+2}, \ldots q_{3N-6}$. But, they may be constants (e.g., if $n = 3N-6$) as happens in the Wilson-Howard treatment of ordinary rigid molecules.

A priori we expect, and experience has confirmed [6, 53)], that the most rapid convergence is obtained if the potential energy is expanded using curvilinear coordinates. However, this advantage is opposed by complications in deriving the kinetic energy. In this respect the rectilinear coordinates are superior. Hoy, Mills and Strey[6)] have discussed these problems in great detail in relation to the treatment of rigid molecules. They show how the nonlinear transformation between rectilinear and curvilinear coordinates can be worked out and employed in expressing the anharmonic force constants corresponding to rectilinear coordinates in terms of purely geometrically defined force constants. It should be emphasized that only force constants of the latter type are isotopically invariant. It is therefore an important advance that such constants can now be used in expressing the spectroscopic constants of the standard treatment.

2.2 Kinetic Energy

The classical Equation for kinetic energy reads

$$T = \frac{1}{2} \sum_{\alpha} m_{\alpha} \dot{\mathbf{R}}_{\alpha} \cdot \dot{\mathbf{R}}_{\alpha} = \frac{1}{2} \sum_{\alpha} m_{\alpha}^{-1} \mathbf{P}_{\alpha} \cdot \mathbf{P}_{\alpha} \tag{2.8}$$

in terms of the mass, m_{α}, and linear velocity, $\dot{\mathbf{R}}_{\alpha}$, or linear momentum, $\mathbf{P}_{\alpha} = m_{\alpha}\dot{\mathbf{R}}_{\alpha}$, of each atom. Usually the rewriting starts by considering the time derivative of Eq. (2.3) which gives the velocity vector for substitution in Eq. (2.8). The three main steps of this standard method are discussed below for comparison with the principles of the alternative procedure based on the momentum transformation.

2.2.1 Usual Treatment

The time derivative of Eq. (2.3) may be written

$$\dot{\mathbf{R}}_{\alpha} = \dot{\mathbf{R}} + \sum_{g} (\dot{\mathbf{e}}_g r_{\alpha g} + \mathbf{e}_g \dot{r}_{\alpha g}) \tag{2.9}$$

By comparison with Eq. (2.6) it is seen that $\dot{r}_{\alpha g}$ depends on the internal velocities exclusively, whereas $\dot{\mathbf{e}}_g$ depends on the rotational velocities. The time derivatives of the Eulerian angles are replaced, however, by the components of the total angular velocity vector, $\boldsymbol{\omega}$, of the molecular system, defined by

$$\dot{\mathbf{e}}_g = \boldsymbol{\omega} \times \mathbf{e}_g \tag{2.10}$$

Substituting this relation into Eq. (2.9) we obtain

$$\dot{\mathbf{R}}_{\alpha} = \dot{\mathbf{R}} + \boldsymbol{\omega} \times \mathbf{r}_{\alpha} + \sum_{k} \frac{\partial \mathbf{r}_{\alpha}}{\partial q_k} \dot{q}_k \tag{2.11}$$

We shall emphasize that this result establishes a linear transformation of the velocities, which can be expressed in matrix form. For this purpose we consider the components of $\dot{\mathbf{R}}_\alpha$ and introduce a common symbol, v_i, $i = 1, 2, \ldots 3N$, for the generalized velocities

$$\{v_1, v_2, \ldots v_{3N}\} = \{\dot{R}_X, \dot{R}_Y, \dot{R}_Z, \omega_x, \omega_y, \omega_z, \dot{q}_1, \dot{q}_2, \ldots \dot{q}_{3N-6}\} \tag{2.12}$$

Further we rewrite

$$\boldsymbol{\omega} \times \mathbf{r}_\alpha = \sum_g (\mathbf{e}_g \omega_g) \times \mathbf{r}_\alpha = \sum_g (\mathbf{e}_g \times \mathbf{r}_\alpha)\, \omega_g \tag{2.13}$$

showing that we generally have

$$\dot{R}_{\alpha F} = \sum_i t_{i,\alpha F}\, v_i \tag{2.14}$$

The transformation coefficients, $t_{i,\alpha F}$, which obviously are functions of the generalized coordinates, may be arranged in a square matrix, $\mathbf{T}$, with columns labelled by i and rows labelled by $\alpha F = 1X, 1Y, 1Z, 2X, 2Y, \ldots NZ$.

In the standard method we proceed by substituting Eq. (2.11) into Eq. (2.8). Here we shall use the equivalent form Eq. (2.14) and obtain

$$T = \frac{1}{2} \sum_{\alpha F} m_\alpha \dot{R}_{\alpha F}^2 = \frac{1}{2} \sum_{ij} K_{ij}\, v_i\, v_j \tag{2.15}$$

where

$$K_{ij} = \sum_{\alpha F} m_\alpha t_{i,\alpha F}\, t_{j,\alpha F} \tag{2.16a}$$

If the elements, K_{ij}, are arranged in a square matrix, $\mathbf{K}$, and the masses in a diagonal matrix, $\mathbf{m}$, where every atomic mass appears three times, we may also obtain the matrix relation

$$\mathbf{K} = \tilde{\mathbf{T}}\, \mathrm{m}\, \mathbf{T} \tag{2.16b}$$

The next step is to introduce generalized and quasi-generalized momenta,

$$\pi_i = \frac{\partial T}{\partial v_i} = \sum_j K_{ij} v_j \tag{2.17}$$

Notice, that we can substitute the individual sums over j in Eq. (2.15) and get

$$T = \frac{1}{2} \sum_i \pi_i v_i \tag{2.18}$$

which is an invariant form, valid for any set of generalized velocities, v_i, with conjugated momenta, π_i.

Since the coordinate transformation Eq. (2.6) must be invertible this also holds for the matrix of first derivatives, **T**, and it is possible to invert **K** as well. We can therefore define the general G-matrix by

$$\mathbf{G} = \mathbf{K}^{-1} \tag{2.19}$$

and express the velocities in terms of the momenta by reversing the transformation [Eq. (2.17)]

$$v_i = \sum_j G_{ij}\,\pi_j \tag{2.20}$$

This is finally substituted into Eq. (2.18) yielding the Hamiltonian form

$$T = \frac{1}{2} \sum_{ij} G_{ij}\,\pi_i\,\pi_j \tag{2.21}$$

We may conclude that the main steps are the formulation of the elements of **T** as functions of the generalized coordinates, the multiplications of Eq. (2.16) and finally the inversion of **K**. From a detailed study of these three steps[1, 36, 43] it is seen that the final inversion is particularly cumbersome, and that the resulting expressions for the G_{ij}-elements may be extremely complicated.

Making a comparison with the alternative procedure sketched in Eqs. (1.1)–(1.3), we see that the first of the main steps has been replaced by an evaluation of s_{ji}-elements, the second step is equivalent and the third step, the inversion, has been avoided. It therefore seems that the alternative is much more straightforward, but this might of course be only an illusion, if the difficulties in evaluating s_{ji}-elements were comparable to those of inverting **K**. However, this is not so, as we shall see below.

2.2.2 Momentum Transformation

In our study of the transformation between Cartesian and generalized momenta we shall start out from the kinetic energy in its invariant form. Using generalized velocities and momenta Eq. (2.18) applies, while the corresponding expression in Cartesian velocities, $\dot{R}_{\alpha F}$, and momenta, $P_{\alpha F} = m_\alpha \dot{R}_{\alpha F}$, reads

$$T = \frac{1}{2} \sum_{\alpha,F} P_{\alpha F} \dot{R}_{\alpha F} \tag{2.22}$$

With a slightly changed notation we can write Eq. (1.1) in the form

$$P_{\alpha F} = \sum_i s_{i,\alpha F}\,\pi_i \tag{2.23}$$

and substituting this as well as Eq. (2.14) into Eq. (2.22) we get

$$T = \frac{1}{2} \sum_{i,j} \sum_{\alpha, F} s_{i,\alpha F} t_{j,\alpha F} \pi_i v_j$$

which because of Eq. (2.18) requires that

$$\sum_{\alpha,F} s_{i,\alpha F} t_{j,\alpha F} = \delta_{ij} \tag{2.24}$$

Hence, the quantities, $s_{i,\alpha F}$, are elements of a square matrix, **S**, inverse of **T**,

$$\mathbf{S}\,\mathbf{T} = \mathbf{T}\,\mathbf{S} = \mathbf{E} \tag{2.25}$$

Recollecting the discussion of the preceding paragraph one might at this point think that the inversion of **K** has been replaced by another inversion, that of **T**, which appears even more complicated, since **T** is unsymmetrical as opposed to **K**. However, an essential advantage of the present method arises from the particular fact that most elements of **S** can be obtained from fundamental properties without considering **T**.

This is realized by reversing the velocity transformation [Eq. (2.14)],

$$v_i = \sum_{\alpha F} s_{i,\alpha F} \dot{R}_{\alpha F} \tag{2.26}$$

If v_i is the time derivative of a generalized coordinate, $v_i = \dot{q}_k$ or $\dot{R}_F$, then the corresponding elements of **S** are defined as the partial derivatives of the coordinate with respect to the Cartesian coordinates. In case of a translational velocity we therefore immediately have

$$s_{F,\alpha F'} = \frac{\partial R_F}{\partial R_{\alpha F'}} = \frac{m_\alpha}{M} \delta_{FF'}, \quad M = \sum_\alpha m_\alpha \tag{2.27}$$

and also for an internal geometrically defined coordinate we can evaluate the partial derivatives,

$$s_{k,\alpha F} = \frac{\partial q_k}{\partial R_{\alpha F}} \tag{2.28}$$

as well-defined functions of the configuration.

We recognize in Eqs. (2.26) and (2.28) a close relation between the **S**-matrix introduced here and the well-known *B*-matrix defined by Wilson, Decius and Cross[1,3]. It must be emphasized, however, that the partial derivatives in Eq. (2.28) should be evaluated with respect to the LS-coordinates and for the instantaneous configuration. They are therefore functions of the generalized coordinates in contrast to the constant *B*-elements.

The rotational velocities, on the other hand, are defined on the basis of a convention for the directions of the molecular axes within the ensemble of atoms. Obviously we are faced here with the quite different problem of determing a relation between the rotational elements of **S** and this axis convention. Before this is discussed in detail we shall introduce a more convenient vector notation.

2.2.3 *s*- and *t*-Vectors

With the basis vectors $\mathbf{e}_F$, $F = X, Y, Z$, and $\mathbf{e}_g$, $g = x, y, z$, of LS and MS we generally write

$$\mathbf{c} = \sum_F \mathbf{e}_F c_F = \sum_g \mathbf{e}_g c_g \tag{2.29}$$

as the relation between a vector **c** of three-dimensional space and its components. This may be used in defining vectors corresponding to the *F*-labelling of **S**- and **T**-elements,

$$\mathbf{s}_{i,\alpha} = \sum_F \mathbf{e}_F s_{i,\alpha F}, \ \mathbf{t}_{j,\alpha} = \sum_F \mathbf{e}_F t_{j,\alpha F} \tag{2.30}$$

These definitions are similar to Wilson's[3)], but more general. The usual *s*-vectors, here written $\mathbf{s}^0_{i,\alpha}$, have special properties because they are formed from derivatives evaluated in the equilibrium configuration. The constant *B*-elements in the treatment of small amplitude vibrations are their components in a molecular system fixed to this equilibrium configuration,

$$B_{k,\alpha g} = s^0_{k,\alpha g} = \left.\frac{\partial q_k}{\partial r_{\alpha g}}\right|_0 \tag{2.31}$$

s-vectors as defined here in a more general sense were used by Meyer and Günthard[54)], also with the purpose of studying unrestricted internal motions. Parts of the present development may be considered as an extension and further generalization of their work.

t-Vectors were introduced by Polo[55)], with the different notation $\boldsymbol{\rho}^0_{i\alpha} \equiv \mathbf{t}^0_{i,\alpha}$, however. Once more the zero indicates that the definition was restricted to *t*-vectors of the equilibrium configuration.

With the present definitions we can write the velocity transformations of Eqs. (2.14) and (2.26) in vector form,

$$\begin{aligned} \dot{\mathbf{R}}_\alpha &= \sum_i \mathbf{t}_{i,\alpha} \mathrm{v}_i; \quad \alpha = 1, 2, \ldots N \\ \mathrm{v}_i &= \sum_\alpha \mathbf{s}_{i,\alpha} \cdot \dot{\mathbf{R}}_\alpha; \quad i = 1, 2, \ldots 3N \end{aligned} \tag{2.32}$$

or, specifying the three types of generalized velocities,

$$\dot{\mathbf{R}}_\alpha = \sum_F \mathbf{t}_{F,\alpha} \dot{R}_F + \sum_g \mathbf{t}_{g,\alpha} \omega_g + \sum_k \mathbf{t}_{k,\alpha} \dot{q}_k \tag{2.33}$$

$$\dot{R}_F = \sum_\alpha s_{F,\alpha} \cdot \dot{\mathbf{R}}_\alpha; \quad F = X, Y, Z$$

$$\omega_g = \sum_\alpha s_{g,\alpha} \cdot \dot{\mathbf{R}}_\alpha; \quad g = x, y, z \tag{2.34}$$

$$\dot{q}_k = \sum_\alpha s_{k,\alpha} \cdot \dot{\mathbf{R}}_\alpha; \quad k = 1, 2, \ldots 3N-6$$

Since ω_F or R_g will never appear in our treatment, confusion should not arise from using the subscripts capital F, small g and k as the only way of characterizing how the s- and t-vectors correspond to the translational, rotational or the vibrational coordinates respectively.

The relation between the **S**- and **T**-matrices [Eq. (2.24)] takes the form

$$\sum_\alpha \mathbf{s}_{i,\alpha} \cdot \mathbf{t}_{j,\alpha} = \delta_{ij} \tag{2.35}$$

and finally the general G-elements, [Eq. (1.3)] are written

$$G_{ij} = \sum_\alpha m_\alpha^{-1} \mathbf{s}_{i,\alpha} \cdot \mathbf{s}_{j,\alpha} \tag{2.36}$$

The appearance of dot products in these expressions means that any G_{ij} is independent of our choice of coordinate system when evaluating s-vector components. As the most directly obtained components we shall prefer those of the molecular system.

2.2.3.1 General Formulae

For the following derivations it may be convenient to recollect a few equations from the algebra of vectors in three-dimensional space:

$$\begin{aligned}
&\mathbf{a} \cdot \mathbf{b} = \mathbf{b} \cdot \mathbf{a} \\
&\mathbf{a} \times \mathbf{b} = -\mathbf{b} \times \mathbf{a} \\
&\mathbf{a} \times \mathbf{b} \cdot \mathbf{c} = \mathbf{b} \times \mathbf{c} \cdot \mathbf{a} = \mathbf{c} \times \mathbf{a} \cdot \mathbf{b} \\
&\mathbf{a} \times (\mathbf{b} \times \mathbf{c}) = \mathbf{b}(\mathbf{a} \cdot \mathbf{c}) - \mathbf{c}(\mathbf{a} \cdot \mathbf{b}) \\
&(\mathbf{a} \times \mathbf{b}) \cdot (\mathbf{c} \times \mathbf{d}) = (\mathbf{a} \cdot \mathbf{c})(\mathbf{b} \cdot \mathbf{d}) - (\mathbf{a} \cdot \mathbf{d})(\mathbf{b} \cdot \mathbf{c})
\end{aligned} \tag{2.37}$$

By comparing Eqs. (2.11), (2.13) and (2.33) it is seen that the three different types of t-vectors are given by

$$\begin{aligned}
&\mathbf{t}_{F,\alpha} = \mathbf{e}_F && \text{(translations)} \\
&\mathbf{t}_{g,\alpha} = \mathbf{e}_g \times \mathbf{r}_\alpha && \text{(rotations)} \\
&\mathbf{t}_{k,\alpha} = \frac{\partial \mathbf{r}_\alpha}{\partial q_k} && \text{(vibrations)}
\end{aligned} \tag{2.38}$$

From Eqs. (2.27) and (2.28) we have

$$\begin{aligned} \mathbf{s}_{F,\alpha} &= \frac{m_\alpha}{M}\,\mathbf{e}_F \qquad \text{(translations)} \\ \mathbf{s}_{k,\alpha} &= \nabla_\alpha q_k \qquad \text{(vibrations)} \end{aligned} \tag{2.39}$$

All components of these vectors can in principle be explicitly written in terms of the generalized coordinates, provided that all atomic coordinate functions, $r_{\alpha g}(q_1, q_2, \ldots q_{3N-6})$, appearing in Eq. (2.6) have been formulated. If the vibrational coordinates are purely geometrically defined, however, the vibrational *s*-vectors and their LS-components, $s_{k,\alpha F}$, are independent of the axis convention used in formulating $r_{\alpha g}$-functions, contrary, of course, to their MS-components, $s_{k,\alpha g}$. Applying Eq. (2.36) we realize that the vibrational part of the *G*-matrix is also independent of the axis convention under these special conditions.

Some fundamental relations involving the vibrational *s*-vectors are independent of the type of internal coordinates. They follow from Eq. (2.35) which particularly implies

$$\sum_\alpha \mathbf{s}_{k,\alpha} \cdot \mathbf{t}_{F,\alpha} = 0, \quad \sum_\alpha \mathbf{s}_{k,\alpha} \cdot \mathbf{t}_{g,\alpha} = 0 \tag{2.40}$$

i.e. three translational and three rotational conditions on each set of vibrational *s*-vectors (given by *k*). The Eq. (2.40) is rewritten, applying Eqs. (2.37) and (2.38),

$$0 = \mathbf{e}_F \cdot \sum_\alpha \mathbf{s}_{k,\alpha}$$

$$0 = \sum_\alpha \mathbf{s}_{k,\alpha} \cdot (\mathbf{e}_g \times \mathbf{r}_\alpha) = \mathbf{e}_g \cdot \sum_\alpha \mathbf{r}_\alpha \times \mathbf{s}_{k,\alpha}$$

which shows that the Eq. (2.40) is equivalent to

$$\sum_\alpha \mathbf{s}_{k,\alpha} = \mathbf{0}, \quad \sum_\alpha \mathbf{r}_\alpha \times \mathbf{s}_{k,\alpha} = \mathbf{0} \tag{2.41}$$

These equations are similar to the conditions formulated by Malhiot and Ferigle[56] for s^0-vectors of the equilibrium configuration. The general conditions were derived earlier[54] by explicitly considering the invariance of vibrational coordinates under translations and rotations.

2.2.3.2 Axis Conventions and Rotational s-Vectors

The translational and rotational conditions [Eq. (2.41)] on vibrational *s*-vectors arose as special cases of Eq. (2.35). This equation also implies similar conditions on vibrational *t*-vectors,

$$\sum_\alpha \mathbf{s}_{F,\alpha} \cdot \mathbf{t}_{k,\alpha} = 0 \tag{2.42}$$

$$\sum_\alpha \mathbf{s}_{g,\alpha} \cdot \mathbf{t}_{k,\alpha} = 0 \tag{2.43}$$

involving the translational and rotational s-vectors.

We shall see that these conditions are closely related to the translational and rotational constraints defining the molecular coordinate system (Sect. 2.1). Furthermore, it turns out that a method for evaluating rotational s-vectors can be based on this relationship. To clarify the principles, we will first discuss the simpler case of translational conditions.

Consider the center of mass conditions [Eq. (2.4)]. The first derivative of the vanishing sums with respect to any generalized coordinate must vanish as well. With an internal coordinate we therefore have

$$\sum_\alpha m_\alpha \frac{\partial \mathbf{r}_\alpha}{\partial q_k} = \mathbf{0} \tag{2.44}$$

However, an exactly equivalent relation arises from Eq. (2.42) when s- and t-vectors are substituted using Eqs. (2.38) and (2.39).

With this observation in mind it immediately seems reasonable that the conditions imposed by the rotational s-vectors [Eq. (2.43)] are equivalent to conditions implied by the convention for orienting the molecular axes.

Exploring the possibilities of this idea we consider the three rotational constraint relations on the atomic coordinates that follow from the axis convention. It is assumed that these relations can be cast in the form

$$C^{(g)}(r_{1x}, r_{1y}, \ldots r_{Nz}) = 0 \tag{2.45}$$

using three functions, labelled by $g = x, y, z$, which vanish for the allowed orientations of the molecular coordinate system. Examples of such functions are given below for the conditions of a principal axis system and for the Eckart conditions[39].

As in the case of the center of mass conditions we can differentiate with respect to a vibrational coordinate. Assuming that any $C^{(g)}$ is a differentiable function of the atomic coordinates we thus obtain

$$0 = \frac{\partial C^{(g)}}{\partial q_k} = \sum_{\alpha g'} \frac{\partial C^{(g)}}{\partial r_{\alpha g'}} \frac{\partial r_{\alpha g'}}{\partial q_k} = \sum_\alpha \mathbf{c}_{g,\alpha} \cdot \mathbf{t}_{k,\alpha} \tag{2.46}$$

Aiming at a relation resembling Eq. (2.43) we have here introduced a new set of vectors given by

$$\mathbf{c}_{g,\alpha} = \sum_{g'} \mathbf{e}_{g'} \frac{\partial C^{(g)}}{\partial r_{\alpha g'}} \tag{2.47}$$

$$g = x, y, z, \quad \alpha = 1, 2, \ldots N$$

Because of their origin they may be called constraint vectors.

We expect that Eqs. (2.43) and (2.46) are equivalent in the sense that they impose equivalent conditions on the vibrational t-vectors. This indicates that we can form the rotational s-vectors of an atom α as linear combinations of the constraint vectors belonging to this atom,

$$\mathbf{s}_{g,\alpha} = \sum_{g'} \eta_{gg'} \, \mathbf{c}_{g',\alpha} \tag{2.48}$$

Moreover, the coefficients $\eta_{gg'}$ must be common to all atoms so that only nine coefficients need to be evaluated.

To prove this and to determine the η-coefficients, it is sufficient to make sure that all conditions of Eq. (2.35) hold with rotational s-vectors given by Eq. (2.48). Since $\mathbf{S}$ is a unique inverse of $\mathbf{T}$ [Eq. (2.25)] only a single unique set of vectors will pass this test.

First Eq. (2.43) holds because of Eq. (2.46).

Secondly we consider the relations involving rotational t-vectors,

$$\begin{aligned} \delta_{gg'} = \sum_{\alpha} \mathbf{s}_{g,\alpha} \cdot \mathbf{t}_{g',\alpha} &= \sum_{\alpha g''} \eta_{gg''} \, \mathbf{c}_{g'',\alpha} \cdot \mathbf{t}_{g',\alpha} \\ &= \sum_{g''} \eta_{gg''} J_{g''g'} \end{aligned} \tag{2.49}$$

where we have introduced the quantities

$$J_{gg'} = \sum_{\alpha} \mathbf{c}_{g,\alpha} \cdot \mathbf{t}_{g',\alpha} \tag{2.50}$$

These, as well as the η-coefficients, can be arranged in 3x3-matrices which obviously must be inverse of each other. Hence, Eq. (2.49) is only fulfilled with a set of η-coefficients uniquely determined as elements forming the inverse of the matrix of J-elements [Eq. (2.50)]. We have hereby established the method of determining the η's which remained up to this stage.

Finally we must control the relations involving the translational t-vectors. This leads to the requirement [compare Eq. (2.41)]

$$\sum_{\alpha} \mathbf{c}_{g,\alpha} = \mathbf{0} \tag{2.51}$$

which is usually fulfilled by inherent properties of the constraint relations [Eq. (2.45)]. However, should problems arise here we only need to add a simple term in order to correct the constraint vectors

$$\mathbf{c}'_{g,\alpha} = \mathbf{c}_{g,\alpha} - \frac{m_\alpha}{M} \sum_{\alpha'} \mathbf{c}_{g,\alpha'} \tag{2.52}$$

The correcting term affects neither Eq. (2.43) nor Eq. (2.50).

Equation (2.48) is perhaps the most important accomplishment of the present paper. Its application in deriving the elements of $\mathbf{G}$ directly [Eq. (2.36)] offers an important advantage compared to the usual method, since it is sufficient to consider the inversion of only a single small (3x3) matrix, $\{J_{gg'}\}$.

In this connection it is particulary interesting to study the general form of the pure rotational part of the G-matrix, usually referred to as the μ-tensor. We find

$$\mu_{gg'''} = \sum_{g'g''} \eta_{gg'} I_{g'g''} \eta_{g''g'''} \tag{2.53}$$

with

$$I_{gg'} = \sum_{\alpha} m_\alpha^{-1} \mathbf{c}_{g,\alpha} \cdot \mathbf{c}_{g',\alpha} \tag{2.54}$$

The symbol I is used for the latter product sum because this quantity may be an inertial tensor element. This turns out to be the case in both examples discussed below.

In terms of 3x3-matrices, $\mathbf{J}$, $\boldsymbol{\eta}$, $\mathbf{I}$ and $\boldsymbol{\mu}$, formed from the elements above, the relations are

$$\boldsymbol{\mu} = \boldsymbol{\eta} \mathbf{I} \boldsymbol{\eta}, \quad \boldsymbol{\eta} = \mathbf{J}^{-1} \tag{2.55}$$

This factorization of the μ-tensor has also been observed in the standard theory of small amplitude motion[57,58], where $\mathbf{I} = \mathbf{I}^0$, the inertial tensor of the equilibrium configuration. Below it is shown how this particular result is following from the Eckart conditions.

2.2.3.2.1 Principal Axis System (PAS). As a first example we shall derive rotational s-vectors for the special case of a PAS, following the procedure outlined above step by step.

First we must formulate the axis convention in accordance with Eq. (2.45). We adopt (f, g, h) as symbols of cyclic permutations of (x, y, z),

$$(f, g, h) = (x, y, z) \quad \text{or} \quad (f, g, h) = (y, z, x) \quad \text{or} \quad (f, g, h) = (z, x, y) \tag{2.56}$$

and write generally

$$C^{(f)} = \sum_{\alpha} m_\alpha r_{\alpha g} r_{\alpha h} = 0 \tag{2.57}$$

With f running over x, y and z all inertial products are constrained to zero, and consequently the molecular axes are principal axes of the instantaneous tensor of inertia.

The constraint vectors are found from Eq. (2.47),

$$\mathbf{c}_{f,\alpha} = m_\alpha(\mathbf{e}_g r_{\alpha h} + \mathbf{e}_h r_{\alpha g}) \tag{2.58}$$

and used in Eqs. (2.50) and (2.54) to express J- and I-elements,

$$J_{ff} = \sum_\alpha \mathbf{c}_{f,\alpha} \cdot (\mathbf{e}_f \times \mathbf{r}_\alpha) = \sum_\alpha (\mathbf{c}_f \times \mathbf{e}_f) \cdot \mathbf{r}_\alpha$$
$$= \sum_\alpha m_\alpha (r_{\alpha g}^2 - r_{\alpha h}^2) = I_h - I_g \tag{2.59}$$
$$J_{fg} = \sum_\alpha \mathbf{c}_{f,\alpha} \times \mathbf{e}_g \cdot \mathbf{r}_\alpha = \sum_\alpha m_\alpha (-r_{\alpha g} r_{\alpha f}) = 0$$

$$I_{ff} = \sum_\alpha m_\alpha (r_{\alpha g}^2 + r_{\alpha h}^2) = I_f$$
$$I_{fg} = \sum_\alpha m_\alpha r_{\alpha g} r_{\alpha f} = 0 \tag{2.60}$$

We have here employed Eq. (2.37) and the vector product rules following from Eq. (2.56),

$$\mathbf{e}_f \times \mathbf{e}_g = \mathbf{e}_h, \quad \mathbf{e}_h \times \mathbf{e}_g = -\mathbf{e}_f \tag{2.61}$$

I_f, I_g and I_h denote the principal moments of inertia and from Eq. (2.60) we notice that **I** of Eq. (2.55) is identical to the instantaneous principal tensor of inertia.

The J-matrix found in Eq. (2.59) is diagonal and is easily inverted, provided the diagonal elements are nonvanishing. We thus obtain a diagonal η-matrix as well, and find s-vectors in agreement with Meyer and Günthard[54],

$$\mathbf{s}_{f,\alpha} = \frac{m_\alpha}{I_h - I_g} (\mathbf{e}_g r_{\alpha h} + \mathbf{e}_h r_{\alpha g}) \tag{2.62}$$

and elements of the diagonal μ-tensor,

$$\mu_{ff} = \frac{I_f}{(I_h - I_g)^2}, \quad \mu_{fg} = 0 \tag{2.63}$$

These results seem quite simple, moreover, the method avoids the use of a reference configuration alleged to present a problem for molecules with large amplitude internal motions[44, 54]. For these reasons the PAS has been used in deriving several Hamiltonians of three atomic molecules[16, 45, 46]. However, it should be emphasized that the PAS may imply very large vibration-rotation coupling terms in case of near symmetric top molecules. This is due to almost vanishing denominators in Eqs. (2.62) as well as (2.63).

As an example consider a planar, near oblate symmetric top molecule. If small amplitude vibrations are assumed, we may expect that the μ-tensor elements are of the order of the reciprocal principal moments of inertia. However, for a planar configuration, with $I_c = I_a + I_b$, it is easily seen that the generally smallest element, μ_{cc}, may reach extreme values when evaluated in a PAS using Eq. (2.63). A numerical example illustrates this:

$$(I_a, I_b, I_c) = (40, 60, 100)\ \mathrm{u\AA}^2$$

corresponds to

$$(\mu_{aa}, \mu_{bb}, \mu_{cc}) = \left(\frac{1}{40}, \frac{1}{60}, \frac{1}{4}\right)\ \mathrm{u}^{-1}\mathrm{\AA}^{-2}$$

Hence, a very large energy contribution from $\mu_{cc}\mathrm{P}_c^2$ must be counterbalanced by coupling terms of the type $G_{kc}p_kP_c$ between vibrational and angular momenta, p_k and P_c.

The assertion that the PAS is convenient for separating rotations and vibrations can be rejected, therefore. We shall see below (Sect. 4) that the small amplitude vibrations are always treated most simply using Eckart conditions, whereas large amplitude motions must be specially taken care of. Principal inertial axes may only be relevant in relation to the reference structure of the Eckart conditions.

2.2.3.2.2 Eckart System (ES). It is well-known how the axis convention proposed by Eckart[39] enters the standard vibration-rotation theory[2, 49–52]. In the alternative method of deriving the kinetic energy the Eckart conditions are used in formulating rotational *s*-vectors. For this purpose we may proceed exactly as in the PAS example above.

The molecular coordinate system of a rigid molecule is defined as the principal axis system of the equilibrium configuration which is taken as as reference. Thus we write

$$\mathbf{r}_\alpha = \mathbf{a}_\alpha + \mathbf{d}_\alpha \tag{2.64}$$

where the set of vectors, $\mathbf{a}_\alpha$; $\alpha = 1, 2, \ldots N$, follows from the orientation of the equilibrium configuration in space, and $\mathbf{d}_\alpha$ is a small displacement vector. The components, $a_{\alpha g}$, are constants, whereas the displacement vector components, $d_{\alpha g}$, depend on the internal coordinates. The position of the reference and the instantaneous configurations relative to each other is defined uniquely by the Eckart conditions,

$$\sum_\alpha m_\alpha \mathbf{r}_\alpha = \sum_\alpha m_\alpha \mathbf{d}_\alpha = \mathbf{0} \tag{2.65}$$

$$\sum_\alpha m_\alpha \mathbf{a}_\alpha \times \mathbf{r}_\alpha = \sum_\alpha m_\alpha \mathbf{a}_\alpha \times \mathbf{d}_\alpha = \mathbf{0} \tag{2.66}$$

Rewritten in terms of vector components Eq. (2.65) gives the three center of mass conditions, while Eq. (2.66) gives the three rotational constraints in a form similar to Eq. (2.45),

$$C^{(g)} = \mathbf{e}_g \cdot \sum_\alpha m_\alpha \mathbf{a}_\alpha \times \mathbf{r}_\alpha = 0; \quad g = x, y, z \tag{2.67}$$

Using this in Eq. (2.47) we then find the constraint vectors,

$$\mathbf{c}_{g,\alpha} = m_\alpha \mathbf{e}_g \times \mathbf{a}_\alpha \tag{2.68}$$

The MS-components of these constraint vectors are constants,

$$c_{g,\alpha f} = m_\alpha a_{\alpha h},\ c_{g,\alpha g} = 0,\ c_{g,\alpha h} = -m_\alpha a_{\alpha f} \tag{2.69}$$

and consequently the elements of **I** [Eq. (2.54)] are constants as well. A closer examination reveals that **I** is simply the inertial tensor of the equilibrium configuration, which we shall assume to be diagonal,

$$I_{gg'} = I^0_{gg'} = \sum_\alpha m_\alpha (\mathbf{e}_g \times \mathbf{a}_\alpha) \cdot (\mathbf{e}_{g'} \times \mathbf{a}_\alpha) \tag{2.70}$$

or

$$\begin{aligned} I_{gg} &= \sum_\alpha m_\alpha (a^2_{\alpha f} + a^2_{\alpha h}) = I^0_g \\ I_{fg} &= -\sum_\alpha m_\alpha a_{\alpha f} a_{\alpha g} = 0 \end{aligned} \tag{2.71}$$

Also the elements of **J**, [Eq. (2.50)], are related to the inertial tensor. Using Eq. (2.38) we thus obtain

$$J_{gg'} = \sum_\alpha m_\alpha (\mathbf{e}_g \times \mathbf{a}_\alpha) \cdot (\mathbf{e}_{g'} \times \mathbf{r}_\alpha) \tag{2.72}$$

which may be rewritten introducing Eq. (2.64)

$$J_{gg'} = I^0_g + \sum_\alpha m_\alpha (\mathbf{e}_g \times \mathbf{a}_\alpha) \cdot (\mathbf{e}_{g'} \times \mathbf{d}_\alpha) \tag{2.73}$$

This important equation deserves several comments. Some will be postponed to subsequent sections, but here we shall first of all notice that $\mathbf{J} = \mathbf{I}^0$ in the equilibrium configuration and that the elements of **J** vary linearly with the atomic displacements. This offers a great advantage when expanding the μ-tensor elements (Sect. 3.2). Furthermore, we can easily show that **J** is a symmetric tensor, although this is not easily recognized from equation (2.73). We use the relations (2.37) in rewriting differences between the off-diagonal elements:

$$J_{gf} - J_{fg} = \sum_\alpha m_\alpha (-a_{\alpha f} d_{\alpha g} + a_{\alpha g} d_{\alpha f}) = -\sum_\alpha m_\alpha \mathbf{e}_h \cdot (\mathbf{a}_\alpha \times \mathbf{d}_\alpha) = 0 \tag{2.74}$$

which vanishes because of equation (2.66).

It is also worth mentioning that **J** is intermediate between $\mathbf{I}^0$ and the instantaneous tensor of inertia $\mathbf{I}^*$. Approximately it holds that

$$\mathbf{J} \cong \frac{1}{2} (\mathbf{I}^0 + \mathbf{I}^*) \tag{2.75}$$

in accordance with the fact that

$$\frac{\partial J_{gg'}}{\partial d_{\alpha g''}} = \frac{1}{2} \frac{\partial I^*_{gg'}}{\partial d_{\alpha g''}} \tag{2.76}$$

2.3 Quantum Mechanical Aspects

Until this stage all discussions have been based on classical mechanics. However, in the present formulation the translation to quantum mechanics is quite straightforward, since quantum momenta can be defined from the momentum transformation with only slight modifications. Furthermore, the resulting expressions are general in the sense that they can be derived without considering any particular representation of the momenta as differential operators. They apply equally well in a wave mechanical context.

2.3.1 Quantum Momenta

In quantum mechanics the momentum transformation [Eq. (2.23)] and its inverse should read

$$\begin{aligned} P_{\alpha F} &= \frac{1}{2} \sum_i (s_{i,\alpha F}\pi_i + \pi_i s_{i,\alpha F}) \\ \pi_i &= \frac{1}{2} \sum_{\alpha F} (t_{i,\alpha F} P_{\alpha F} + P_{\alpha F} t_{i,\alpha F}) \end{aligned} \tag{2.77}$$

This holds generally, even in cases where some of the generalized momenta, π_i, are "quasi-momenta"[59], i.e. momenta not conjugated to generalized coordinates, e.g. the angular momenta.

Generalized momentum operators as defined by Eq. (2.77) can be used in wave mechanical as well as in matrix mechanical formulations. It ensures that the operators are Hermitian, and that momenta, π_i, conjugated to generalized coordinates, q_i, fulfil commutation relations similar to the canonical relations of Cartesian coordinates and momenta,

$$[\pi_i, q_j] = -i\hbar\delta_{ij}, \quad [\pi_i, \pi_j] = 0, \quad [q_i, q_j] = 0 \tag{2.78}$$

2.3.2 Quantum Kinetic Energy

The classical kinetic energy expression in terms of Cartesian momenta,

$$T = \frac{1}{2} \sum_{\alpha F} m_\alpha^{-1} P_{\alpha F}^2 \tag{2.79}$$

can be directly transferred to quantum mechanics. The quantum kinetic energy expressed by generalized momenta and coordinates may therefore be similarly derived applying the momentum transformation [Eq. (2.77)]. The equation may be rewritten in two alternative forms,

$$P_{\alpha F} = \sum_i \left(\pi_i s_{i,\alpha F} - \frac{1}{2}\left[\pi_i, s_{i,\alpha F}\right]\right)$$
$$= \sum_i \left(s_{i,\alpha F}\, \pi_i + \frac{1}{2}\left[\pi_i, s_{i,\alpha F}\right]\right) \qquad (2.80)$$

Substitution into Eq. (2.79) thus yields

$$T = \frac{1}{2} \sum_{i,j} \pi_i G_{ij} \pi_j + U \qquad (2.81)$$

which is almost identical to the classical expression [Eq. (2.21)]. It must be emphasized, however, that the sequence of operators in the sum is crucial and that a mass dependent term, U, contributes to the potential energy,

$$U = \frac{1}{4} \sum_{ij,\alpha F} m_\alpha^{-1} (s_{i,\alpha F}[\pi_i, [\pi_j, s_{j,\alpha F}]] + \frac{1}{2} [\pi_i, s_{i,\alpha F}][\pi_j, s_{j,\alpha F}]) \qquad (2.82)$$

For rigid molecules this expression of U reduces in agreement with the result obtained by Watson[58, 59] from another starting point, namely Podolsky's equation[60]. In some cases, however, Eq. (2.82) may offer the advantage that it also applies when quasi-momenta are used.

The expression of U may be rewritten using $[\pi_i, \mathbf{e}_F] = \mathbf{0}$ and taking the general properties of translational and angular momenta into account. We can separate the set of momenta, π_i, into three types just as the velocities v_i were separated [Eq. (2.12)].

$$\{\pi_1, \pi_2, \ldots \pi_{3N}\} = \{\mathscr{P}_X, \mathscr{P}_Y, \mathscr{P}_Z, P_x, P_y, P_z, p_1, p_2, \ldots p_{3N-6}\} \qquad (2.83)$$

where $\mathscr{P}_F$ is a translational, P_g is a rotational and p_k is a vibrational momentum. We also generally have the commutation relations,

$$[\mathscr{P}_F, \mathbf{s}_{F,\alpha}] = \mathbf{0}, \quad [P_g, \mathbf{s}_{i,\alpha}] = -\mathrm{i}\,\hbar\, \mathbf{e}_g \times \mathbf{s}_{i,\alpha} \qquad (2.84)$$

where the last equation follows from the commutation relations of angular momenta with directional cosines[61]. Therefore all terms involving translations will vanish and the remaining may conveniently be divided into pure rotational, pure vibrational and mixed contributions.

Thus, in the rotational term i, j of Eq. (2.82) run over x, y and z. Using Eq. (2.84) it is reduced to

$$U_1 = \frac{\hbar^2}{8} \sum_{\alpha g g'} m_\alpha^{-1} (\mathbf{e}_g \times \mathbf{s}_{g,\alpha}) \cdot (\mathbf{e}_{g'} \times \mathbf{s}_{g',\alpha}) \qquad (2.85)$$

The vibration-rotation term similarly becomes

$$U_2 = \frac{i\hbar}{4} \sum_{\alpha g k} m_\alpha^{-1} (\mathbf{e}_g \times \mathbf{s}_{k,\alpha}) \cdot [p_k, \mathbf{s}_{g,\alpha}] \qquad (2.86)$$

whereas the pure vibrational term cannot be reduced without special assumptions as to the nature of the vibrational coordinates,

$$U_3 = \frac{1}{4} \sum_{\alpha k k'} m_\alpha^{-1} (\mathrm{s}_{k,\alpha} \cdot [p_k, [p_{k'}, \mathrm{s}_{k',\alpha}]] + \frac{1}{2} [p_k, \mathrm{s}_{k,\alpha}] \cdot [p_{k'}, \mathrm{s}_{k',\alpha}]) \quad (2.87)$$

Alternatively,

$$U_3 = \frac{1}{8} \sum_{kk'} ([p_k, [p_{k'}, G_{kk'}]] - \sum_\alpha [p_k, \mathrm{s}_{k',\alpha}] \cdot [p_{k'}, \mathrm{s}_{k,\alpha}] m_\alpha^{-1}) \quad (2.88)$$

3 Hamiltonian of Rigid Molecules

In this section we shall see how the principles outlined above are applied to evaluate the Wilson-Howard Hamiltonian[1, 2]. However, most of the derivation may be worked out without explicitly assuming that rectilinear internal coordinates are used. We shall take advantage of this in that we will also examine the general consequences of the Eckart conditions as opposed to the special properties connected with the introduction of linearized coordinates. As an intermediate result we will therefore obtain a Hamiltonian which is exactly equivalent to the one which Quade derived for the case of geometrically defined curvilinear coordinates[7].

For the present linear molecules are excluded from the treatment. Their special problems are discussed at the end of this section.

3.1 Coordinate Transformations

The first step is to formulate the relationship between Cartesian displacement coordinates, $d_{\alpha g}$, and internal displacement coordinates, q_k; $k = 1, 2, \ldots 3N-6$. For rigid molecules undergoing small amplitude vibrations we can assume that an expansion from the equilibrium configuration,

$$q_k = \sum_{\alpha g} \frac{\partial q_k}{\partial r_{\alpha g}}\bigg|_0 d_{\alpha g} + \frac{1}{2} \sum_{\alpha\alpha' g g'} \frac{\partial^2 q_k}{\partial r_{\alpha g} \partial r_{\alpha' g'}}\bigg|_0 d_{\alpha g} d_{\alpha' g'} + \ldots \quad (3.1)$$

will converge rapidly. The inverse transformation reads

$$d_{\alpha g} = \sum_k \frac{\partial r_{\alpha g}}{\partial q_k}\bigg|_0 q_k + \frac{1}{2} \sum_{kk'} \frac{\partial^2 r_{\alpha g}}{\partial q_k \partial q_{k'}}\bigg|_0 q_k q_{k'} + \ldots \quad (3.2)$$

By the definitions of Eqs. (2.38)–(2.39) the partial derivatives of these two equations are the same as those appearing in expansions of *s*- and *t*-vector components,

$$s_{k,\alpha g} = \left.\frac{\partial q_k}{\partial r_{\alpha g}}\right|_0 + \sum_{\alpha' g'} \left.\frac{\partial^2 q_k}{\partial r_{\alpha g}\partial r_{\alpha' g'}}\right|_0 d_{\alpha' g'} + \ldots \tag{3.3}$$

$$t_{k,\alpha g} = \left.\frac{\partial r_{\alpha g}}{\partial q_k}\right|_0 + \sum_{k'} \left.\frac{\partial^2 r_{\alpha g}}{\partial q_k \partial q_{k'}}\right|_0 q_{k'} + \ldots \tag{3.4}$$

Alternatively, the coordinate transformations may therefore be written

$$q_k = \sum_{\alpha g} s^0_{k,\alpha g}\, d_{\alpha g} + \frac{1}{2} \sum_{\alpha\alpha' g g'} \left.\frac{\partial s_{k,\alpha g}}{\partial r_{\alpha' g'}}\right|_0 d_{\alpha g}\, d_{\alpha' g'} + \ldots \tag{3.5}$$

$$d_{\alpha,g} = \sum_k t^0_{k,\alpha g}\, q_k + \frac{1}{2} \sum_{kk'} \left.\frac{\partial t_{k,\alpha g}}{\partial q_{k'}}\right|_0 q_k q_{k'} + \ldots \tag{3.6}$$

In a general treatment allowing for curvilinear valence coordinates we must evaluate derivatives of bond lengths and angles for use in Eq. (3.1). Aiming at a Hamiltonian correct to the second order up to third order derivatives are required[6]. Next the derivatives of Eq. (3.2) must be found by interverting Eq. (3.1) with the displacements subjects to Eckart's conditions (here second order suffices). The complexity of solving this problem is an almost insuperable barrier to the practical use of curvilinear coordinates[62]. Rectilinear coordinates are therefore usually introduced as discussed previously (Sect. 2.1).

3.1.1 Rectilinear Coordinates

These may be related to and named as valence coordinates if the same first order derivatives, $s^0_{k,\alpha g}$ and $t^0_{k,\alpha g}$, apply to both classes of coordinates. A linearized coordinate is defined from its counterpart among the valence coordinates by truncating the expansions [Eqs. (3.4) and (3.6)] after the first term. This means that

$$\mathbf{d}_\alpha = \sum_k \mathbf{t}^0_{k,\alpha}\, q_k \tag{3.7}$$

But, it should immediately be emphasized that it is not allowed to treat the expansions [Eqs. (3.3) and (3.5)] in a similar way. This is due to the fact that the vibrational s-vectors are subject to the generalized Malhiot-Ferigle conditions [Eq. (2.41)].

The origin of these relations was Eq. (2.40) and their implications are most easily studied from this starting point. Hence, we find, using Eqs. (2.38), (2.64), (3.7) and

$$\mathbf{t}^0_{g,\alpha} = \mathbf{e}_g \times \mathbf{a}_\alpha \tag{3.8}$$

that

$$\begin{aligned}\sum_\alpha \mathbf{s}^0_{k,\alpha} \cdot \mathbf{t}_{g,\alpha} &= \sum_\alpha \mathbf{s}^0_{k,\alpha} \cdot (\mathbf{t}^0_{g,\alpha} + \mathbf{e}_g \times \mathbf{d}_\alpha) \\ &= \sum_{\alpha k'} (\mathbf{e}_g \times \mathbf{t}^0_{k',\alpha}) \cdot \mathbf{s}^0_{k,\alpha}\, q_{k'}\end{aligned} \tag{3.9}$$

This sum is generally nonvanishing.

The present stage is suitable for the introduction of the Coriolis coupling constants, $\zeta^g_{kk'}$ [63, 64]. This may seem curious, but it is in accordance with the fact that these constants are appropriate only when rectilinear coordinates are involved. This will be further discussed below in relation to the vibration-rotation part of the G-matrix. Here it is convenient to give a general definition,

$$\zeta^g_{k'k} = \sum_\alpha \mathbf{e}_g \cdot (\mathbf{t}^0_{k',\alpha} \times \mathbf{s}^0_{k,\alpha}) \tag{3.10}$$

which applies to any set of rectilinear coordinates (not normal coordinates only). Equation (3.10) is used in rewritting Eq. (3.9),

$$\sum_\alpha \mathbf{s}^0_{k,\alpha} \cdot \mathbf{t}_{g,\alpha} = \sum_{k'} q_{k'} \zeta^g_{k'k} \tag{3.11}$$

and finally we realize that vibrational s-vectors, which fulfil equation (2.40) and all other conditions included in equation (2.35), may be constructed by adding small contributions from rotational s-vectors,

$$\mathbf{s}_{k,\alpha} = \mathbf{s}^0_{k,\alpha} - \sum_{gk'} q_{k'} \zeta^g_{k'k} \mathbf{s}_{g,\alpha} \tag{3.12}$$

3.2 μ-Tensor

Once again returning to a general set of vibrational coordinates we shall now study the pure rotational part of the kinetic energy, given by

$$T_{\text{rot}} = \frac{1}{2} \sum_{gg'} \mu_{gg'} P_g P_{g'} \tag{3.13}$$

where the μ-elements can be expaned on the basis of Eqs. (2.55), (2.70)–(2.73) and (3.6). Defining

$$\begin{aligned} \mathbf{J}' &= \mathbf{J} - \mathbf{I}^0 \\ J'_{gg'} &= \sum_\alpha m_\alpha (\mathbf{e}_g \times \mathbf{a}_\alpha) \cdot (\mathbf{e}_{g'} \times \mathbf{d}_\alpha) \end{aligned} \tag{3.14}$$

we may express $\boldsymbol{\eta}$ by

$$\boldsymbol{\eta} = \sum_{n=0}^{\infty} \boldsymbol{\mu}^0 (-\mathbf{J}' \boldsymbol{\mu}^0)^n, \; \boldsymbol{\mu}^0 = (\mathbf{I}^0)^{-1} \tag{3.15}$$

yielding,

$$\boldsymbol{\mu} = \sum_{n=0}^{\infty} (n+1) \boldsymbol{\mu}^0 (-\mathbf{J}' \boldsymbol{\mu}^0)^n \tag{3.16}$$

Truncated after the second order term we have

$$\boldsymbol{\mu} = \boldsymbol{\mu}^0 - 2\,\boldsymbol{\mu}^0\,\mathbf{J}'\,\boldsymbol{\mu}^0 + 3\,\boldsymbol{\mu}^0\,\mathbf{J}'\,\boldsymbol{\mu}^0\,\mathbf{J}'\,\boldsymbol{\mu}^0 \tag{3.17}$$

which compares with the expansion given by Watson[58].

Partial derivatives of $\boldsymbol{\mu}$ with respect to internal coordinates of any type are then obtained by first writing Eq. (3.6) in vector form

$$\mathbf{d}_\alpha = \sum_k \mathbf{t}^0_{k,\alpha}\, q_k + \frac{1}{2} \sum_{kk'} \mathbf{t}^0_{kk',\alpha}\, q_k q_{k'} \ldots \tag{3.18}$$

with

$$\mathbf{t}^0_{kk'} = \left.\frac{\partial^2 \mathbf{r}_\alpha}{\partial q_k\, \partial q_{k'}}\right|_0 \tag{3.19}$$

Further we obtain the derivatives of $J_{gg'}$,

$$\begin{aligned} J^{(k)}_{gg'} &= \sum_\alpha m_\alpha (\mathbf{e}_g \times \mathbf{a}_\alpha) \cdot (\mathbf{e}_{g'} \times \mathbf{t}^0_{k,\alpha}) \\ J^{(kk')}_{gg'} &= \sum_\alpha m_\alpha (\mathbf{e}_g \times \mathbf{a}_\alpha) \cdot (\mathbf{e}_{g'} \times \mathbf{t}^0_{kk',\alpha}) \end{aligned} \tag{3.20}$$

which finally are substituted into Eq. (3.17) giving

$$\begin{aligned} \mu^{(k)}_{gg'} &= -2\,\mu^0_{gg}\, J^{(k)}_{gg'}\, \mu^0_{g'g'} \\ \mu^{(kk')}_{gg'} &= -2\,\mu^0_{gg}\, J^{(kk')}_{gg'}\, \mu^0_{g'g'} \\ &\quad + \frac{3}{4} \sum_{g''} I^0_{g''} \left(\mu^{(k)}_{gg''}\, \mu^{(k')}_{g''g'} + \mu^{(k')}_{gg''}\, \mu^{(k)}_{g''g'}\right) \end{aligned} \tag{3.21}$$

Terms in $J^{(kk')}_{gg'}$ vanish if linearized coordinates are assumed.

Notice, that $J^{(kk')}_{gg'}$ is not related in any simple way to second derivatives of the instantaneous tensor of inertia, $\mathbf{I}^*$,

$$\begin{aligned} \left.\frac{\partial^2 I^*_{gg'}}{\partial q_k \partial q_{k'}}\right|_0 &= 2\,A^{(gg')}_{kk'} = 2\,J^{(kk')}_{gg'} + \\ &\sum_\alpha m_\alpha \{(\mathbf{e}_g \times \mathbf{t}^0_{k,\alpha}) \cdot (\mathbf{e}_{g'} \times \mathbf{t}^0_{k',\alpha}) + (\mathbf{e}_g \times \mathbf{t}^0_{k',\alpha}) \cdot (\mathbf{e}_{g'} \times \mathbf{t}^0_{k,\alpha})\} \end{aligned} \tag{3.22}$$

contrary to the simple relationship for the first derivatives,

$$\frac{\partial I^*_{gg'}}{\partial q_k} = a^{(gg')}_k = 2\,J^{(k)}_{gg'} \tag{3.23}$$

This impedes a detailed comparison with the expansion given by Quade[7]. The expressions above seem simpler, however, indicating that the present method is more convenient. Moreover, some simplifying relations for planar molecules are easily derived in this connection.

3.2.1 Planar Molecules

Labelling the MS-axes by *a, b, c* according to increasing moments of inertia we can assume that all *c*-coordinates of the reference configuration are zero, $a_{\alpha,c} = 0$; $\alpha = 1, 2 \ldots N$. From Eq. (3.14) it then follows that

$$J_{ac} = J_{ca} = J_{bc} = J_{cb} = 0 \tag{3.24}$$

and we see that **J** is partitioned into a 2 x 2 diagonal block and a single diagonal element, J_{cc}. This block form is preserved by inversion and by multiplications in Eq. (2.55), so we conclude that μ has this block form as well,

$$\mu_{ac} = \mu_{ca} = \mu_{bc} = \mu_{cb} = 0 \tag{3.25}$$

Notice, that this was derived without any assumptions as to type of internal coordinates, curvi- or rectilinear.

From Eq. (3.14) we may also deduce the general relation

$$J_{aa} + J_{bb} = J_{cc} \tag{3.26}$$

similar to the relation for moments of inertia of a planar configuration. However, Eq. (3.26) applies to an arbitrary configuration as long as the reference is planar, in accordance with the fact that all *J*-elements are independent of atomic displacements in the direction of the *c*-axis,

$$\frac{\partial J_{gg'}}{\partial d_{\alpha c}} = 0 \tag{3.27}$$

The importance of these results in relation to centrifugal distortion is well understood[65].

3.3 Coriolis Coupling

The part of the kinetic energy involving both vibrational and angular momenta is

$$T_{\text{vib-rot}} = \sum_{gk} G_{gk} P_g p_k; \; g = x, y, z; k = 1, 2 \ldots 3N-6 \tag{3.28}$$

where we obtain the *G*-elements using Eqs. (2.36), (2.48) and (2.68),

$$G_{gk} = \sum_{\alpha g'} \eta_{gg'} (\mathbf{e}_{g'} \times \mathbf{a}_\alpha) \cdot \mathbf{s}_{k,\alpha} \tag{3.29}$$

These elements vanish in the equilibrium configuration. In order to prove this, we may substitute the cross products, [Eq. (3.8)], giving

$$G_{gk} = \sum_{\alpha g'} \eta_{gg'} \mathbf{t}^0_{g',\alpha} \cdot \mathbf{s}_{k,\alpha} \tag{3.30}$$

In the equilibrium configuration all $\mathbf{s}_{k,\alpha} = \mathbf{s}^0_{k,\alpha}$ and Eq. (2.40) applies. Alternatively we may substitute $\mathbf{a}_\alpha$ by $\mathbf{r}_\alpha - \mathbf{d}_\alpha$ [Eq. (2.64)], and using Eqs. (2.38) and (2.40) we obtain

$$G_{gk} = -\sum_{\alpha g'} \eta_{gg'} (\mathbf{e}_{g'} \times \mathbf{d}_\alpha) \cdot \mathbf{s}_{k,\alpha} \tag{3.31}$$

which vanish when all displacements are zero, $\mathbf{d}_\alpha = \mathbf{0}$.

The Eckart conditions play an important role in this connection. We shall discuss this in more detail below, since the arguments presented apply equally well to the treatment of nonrigid molecules. Hence, to study the basis of introducing Eckart conditions, let us for a moment go back to an earlier stage where axis conventions were not yet formulated. We recapitulate that we are looking for the conditions required in order that the atomic position coordinates, $r_{\alpha g}$, can be given as unique functions of $3N-6$ internal coordinates, or equivalently stated, in order that the expansion [Eq. (3.6)] can be determined as a unique inverse of Eq. (3.5).

Aiming at a separate treatment of vibration and rotation it is obvious that we must look for such axis orientations where the Coriolis coupling term [Eq. (3.28)] either vanishes or at least can be treated as a small perturbation. With Eq. (2.48) we can generally write

$$G_{gk} = \sum_{\alpha g'} \eta_{gg'} m_\alpha^{-1} \mathbf{c}_{g',\alpha} \cdot \mathbf{s}_{k,\alpha} \tag{3.32}$$

Recollecting Eq. (2.40) it is natural to check if axial constraints could be so formulated that differentiation results in equations of the form used in Eq. (2.46) with constraint vectors given by

$$\mathbf{c}_{g,\alpha} = m_\alpha \mathbf{t}_{g,\alpha} \tag{3.33}$$

Were this possible, we would always have $G_{gk} = 0$. We therefore consider the reverse process, the integration of Eq. (2.46), and ask whether it is possible to solve the three integral equations,

$$\int \sum_{\alpha k} \mathbf{c}_{g,\alpha} \cdot \frac{\partial \mathbf{r}_\alpha}{\partial q_k} \mathrm{d}q_k = \int \sum_\alpha \mathbf{c}_{g,\alpha} \cdot \mathrm{d}\mathbf{r}_\alpha = 0, \quad g = x, y, z \tag{3.34}$$

With constraint vectors given by Eq. (3.33) the answer is no[54]. We cannot totally eliminate Coriolis coupling, but, on the other hand, this is not too serious, since vibration-rotation coupling is unavoidable anyway because of the dependence of μ-elements on vibrational coordinates.

However, with constraint vectors that are independent of the internal coordinates an integration can be performed. Furthermore, using

$$\mathbf{c}_{g,\alpha} = m_\alpha \mathbf{t}^0_{g,\alpha} \tag{3.35}$$

we usually achieve orders of magnitude for the two sources of vibration-rotation coupling which are equal and minimal. Integration [Eq. (3.34)] yields the constraint relations,

$$\sum_\alpha m_\alpha \mathbf{t}^0_{g,\alpha} \cdot \mathbf{r}_\alpha = 0, \quad g = x, y, z \tag{3.36}$$

which are the Eckart conditions.

When linearized coordinates are considered, an additional reason arises for requiring structurally independent constraint vectors. This was also noticed by Eckart[39], but, in the present context, it is most easily understood by considering that Eq. (2.46) requires constant constraint vector components, $c_{g,\alpha g'}$, since the t-vector components, $t_{k,\alpha g} = t^0_{k,\alpha g}$, are constants. Hence, linearized internal coordinates are incompatible with constraint vectors like those of the PAS (Sect. 2.2.3.2.1).

Introducing the particular s-vectors of linearized coordinates [Eq. (3.12)] we obtain a special form of G_{gk}, [Eq. (3.30)],

$$G_{gk} = -\sum_{g'k'} \mu_{gg'} q_{k'} \zeta^{g'}_{k'k} \tag{3.37}$$

where Eqs. (3.35), (2.48) and (2.36) were employed in rewriting. The appearance of μ-elements in this equation makes it appropriate to introduce the vibrational angular momenta,

$$p\!\!\!/_g = \sum_{kk'} q_k \zeta^g_{kk'} p_{k'} \tag{3.38}$$

so that the coupling term [Eq. (3.28)] becomes the familiar

$$T_{\text{vib-rot}} = -\sum_{gg'} \mu_{gg'} P_g \, p\!\!\!/_{g'} \tag{3.39}$$

But it is emphasized that this is another particular consequence of using rectilinear coordinates.

In the general expressions of Eqs. (3.30) or (3.31), there are no reasons for introducing μ- or ζ-functions. When expanding, however, the leading term in first power of the coordinates can be written with coefficients,

$$G^{(k')}_{gk} = -\mu^0_{gg} \zeta^g_{k'k} \tag{3.40}$$

applicable when curvi- as well as rectilinear coordinates are used. Therefore, the class of coordinates is unimportant for the Coriolis coupling effects as long as first order perturbation terms are concerned.

3.4 Vibrational *G*-Matrix

The vibrational elements are given by

$$G_{kk'} = \sum_{\alpha} m_{\alpha}^{-1} \mathbf{s}_{k,\alpha} \cdot \mathbf{s}_{k',\alpha} \tag{3.41}$$

which can be expanded using Eq. (3.3). In the special case of linearized coordinates Eq. (3.12) applies.

If for a given molecule we consider sets of curvilinear and rectilinear coordinates that are interrelated in the sense discussed in Sect. 3.1.1, we will obtain identical leading terms in corresponding expansions, irrespective of the class of coordinates. Since the harmonic part of the potential function is also independent of whether we use curvi- or rectilinear coordinates[6], we can use the same linear transformation to normal coordinates as well. In both cases the transformation matrix, **L**, is determined by the set of equations,

$$\mathbf{L}\,\mathbf{L}^{\dagger} = \mathbf{G}^{0}, \quad \mathbf{L}^{\dagger}\,\mathbf{F}\,\mathbf{L} = \mathbf{\Lambda} \tag{3.42}$$

appearing in the well-known GF-method[2].

From the *L*-matrix and its inverse it is possible to calculate *s*- and *t*-vectors corresponding to normal coordinates, $Q_1, Q_2, \ldots Q_{3N-6}$. All previously derived formulae apply equally well to these transformed vectors. But, when normal coordinates are considered, the zeroth order vibrational *G*-matrix becomes a unit matrix and the vectors present some special properties,

$$\begin{aligned} G^{0}_{kk'} &= \sum_{\alpha} m_{\alpha}^{-1} \mathbf{s}^{0}_{k,\alpha} \cdot \mathbf{s}^{0}_{k',\alpha} = \delta_{kk'} \\ &\sum_{\alpha} m_{\alpha}\, \mathbf{t}^{0}_{k,\alpha} \cdot \mathbf{t}^{0}_{k',\alpha} = \delta_{kk'} \end{aligned} \tag{3.43}$$

Considering Eq. (2.35) we find the relation

$$\mathbf{s}^{0}_{k,\alpha} = m_{\alpha} \mathbf{t}^{0}_{k,\alpha} \tag{3.44}$$

which further suggests the introduction of *l*-vectors[63],

$$\mathbf{s}^{0}_{k,\alpha} = m_{\alpha}^{1/2}\, \mathbf{l}_{k,\alpha}, \quad \mathbf{t}^{0}_{k,\alpha} = m_{\alpha}^{-1/2}\, \mathbf{l}_{k,\alpha} \tag{3.45}$$

$$\sum_{\alpha} \mathbf{l}_{k,\alpha} \cdot \mathbf{l}_{k',\alpha} = \delta_{kk'} \tag{3.46}$$

3.5 Wilson-Howard Hamiltonian

Assuming linearized normal coordinates we can substitute Eq. (3.45) into previously derived equations. Thus Eq. (3.10) yields

$$\zeta_{kk'}^{g} = \sum_{\alpha} \mathbf{e}_g \cdot (\mathbf{l}_{k,\alpha} \times \mathbf{l}_{k',\alpha}) = -\zeta_{k'k}^{g} \tag{3.47}$$

in agreement with the original definitions[63, 64].

The Hamiltonian form of the kinetic energy is most easily obtained by considering the momentum transformation directly,

$$\mathbf{P}_\alpha = \sum_F \mathbf{s}_{F,\alpha}\, \mathscr{P}_F + \sum_g \mathbf{s}_{g,\alpha}\, P_g + \sum_k \mathbf{s}_{k,\alpha}\, p_k \tag{3.48}$$

which is rewritten employing Eqs. (3.12) and (3.38),

$$\mathbf{P}_\alpha = \sum_F \mathbf{s}_{F,\alpha}\, \mathscr{P}_F + \sum_g \mathbf{s}_{g,\alpha} (P_g - p\!\!/_g) + \sum_k \mathbf{s}_{k,\alpha}^0\, p_k \tag{3.49}$$

$P_{\alpha F}$ is then substituted in the basic kinetic energy [Eq. (2.79)], recollecting Eqs. (2.36), (2.39) and (3.45) and immediately we obtain

$$2\,T = \frac{1}{M} \sum_F \mathscr{P}_F^2 + \sum_{gg'} \mu_{gg'} (P_g - p\!\!/_g)(P_{g'} - p\!\!/_{g'}) + \sum_k p_k^2 \tag{3.50}$$

in agreement with Wilson and Howard[1].

The only functions for which expansions must be considered are $\mu_{gg'}$ and the potential energy. We recall that μ is given by Eqs. (3.16) or (3.17), and that J'-elements are linear in the rectilinear normal coordinates, with

$$J_{gg'}^{(k)} = \frac{1}{2}\, a_k^{(gg')} = \sum_\alpha m_\alpha^{1/2} (\mathbf{e}_g \times \mathbf{a}_\alpha) \cdot (\mathbf{e}_{g'} \times \mathbf{l}_{k,\alpha}) \tag{3.51}$$

The transformation to a quantum mechanical Hamilton operator (Sect. 2.3) has been discussed by Watson[58]. The operator resulting from the kinetic energy in Eq. (3.50), omitting the translational energy, is given by

$$H = \frac{1}{2} \sum_{gg'} (P_g - p\!\!/_g)\, \mu_{gg'} (P_{g'} - p\!\!/_{g'}) + \frac{1}{2} \sum_k p_k^2 + U + V \tag{3.52}$$

with

$$U = -\frac{1}{8}\, \hbar^2 \sum_g \mu_{gg} \tag{3.53}$$

3.6 Linear Molecules

The method of treating nonlinear molecules is applicable to linear molecules as well with only a few special properties to consider.

The particular problems of linear molecules are related to the choice of generalized coordinates. It is usual to define $3N-5$ internal coordinates, whereas there seems to be some obscurity in the literature as to the precise definition of the molecular coordinate system and the rotational coordinates. A recent discussion by Watson[59] still leaves a choice open which seems unnecessary in the present context. A unique definition of the molecular system involving only two polar angles is suggested below.

A quite different problem connected with the use of $3N-5$ internal coordinates and only two rotational coordinates should also be emphasized here. This choice of generalized coordinates requires that rectilinear coordinates must be used in all cases except for three-atom molecules. In general curvilinear coordinates can only be introduced by the special method presented in Sect. 4, using $3N-6$ vibrational and the three usual Euler angles, whereas the treatment suggested by Quade[7] is applicable to three-atom molecules only. The reason for this special restriction is that the planes formed by bending a linear molecule will not in general contain the molecular z-axis as determined from the Eckart conditions. Therefore we cannot define true valence coordinates in pairs

$$\mathcal{R}_{i1} = \sin\rho_i \cos\chi_i, \quad \mathcal{R}_{i2} = \sin\rho_i \sin\chi_i \tag{3.54}$$

where ρ_i is the angle between two adjacent bonds and χ_i should (but cannot) define the orientation of the normal to the plane of the bonds[6,7].

In the following we therefore consider only a treatment based on rectilinear internal coordinates, of which $N-1$ are related to the stretchings and $2N-4$ are related in pairs to angle bendings. These coordinates determine the positions of atoms relative to a molecular coordinate system which can be rotated in space only by varying two of the Euler angles, φ and θ, whereas the third is kept constant, $\chi = 0$ (see Ref. 2, App. I). Thus the y-axis of MS always stays within the XY-plane of LS, i.e. $\Phi_{Zy} = 0$.

The z-axis is chosen as the axis of the linear reference configuration (i.e. $a_{\alpha x} = a_{\alpha y} = 0$, all α) and Eq. (2.66) gives rise to only two non-trivial constraints on the displacements,

$$\begin{aligned} &\sum_\alpha m_\alpha(\mathbf{e}_x \times \mathbf{a}_\alpha)\cdot\mathbf{d}_\alpha = 0 \quad \text{or} \quad \sum_\alpha m_\alpha a_{\alpha z} d_{\alpha y} = 0 \\ &\sum_\alpha m_\alpha(\mathbf{e}_y \times \mathbf{a}_\alpha)\cdot\mathbf{d}_\alpha = 0 \quad \text{or} \quad \sum_\alpha m_\alpha a_{\alpha z} d_{\alpha x} = 0 \end{aligned} \tag{3.55}$$

Vectors $\mathbf{e}_z \times \mathbf{a}_\alpha$ vanish for all α. However, Eq. (3.55) is sufficient to determine the orientation of the z-axis and therefore the corresponding polar angles, φ and θ, can be determined as well.

The coordinate transformation can now be written in a form similar to Eq. (2.6),

$$R_{\alpha F} = R_F + \sum_g \Phi_{Fg}(\varphi,\theta)\,(a_{\alpha g} + d_{\alpha g}(q_1, q_2, \ldots q_{3N-5})) \tag{3.56}$$

But the velocity transformation cannot be evaluated by the procedure used in Sect. 2.2.1. An equation similar to Eq. (2.33) can be obtained, however, if we use the rotational velocities $\dot\varphi$ and $\dot\theta$,

$$\dot{R}_{\alpha F} = \sum_{F'} t_{F',\alpha F} \dot{R}_{F'} + t_{\varphi,\alpha F}\dot{\varphi} + t_{\theta,\alpha F}\,\dot{\theta} + \sum_{k} t^{0}_{k,\alpha F}\,\dot{q}_k \tag{3.57}$$

The t-vectors of φ and θ are given in analogy with the t-vectors of Eq. (2.38) by

$$\mathbf{t}_{\varphi,\alpha} = \mathbf{e}_Z \times \mathbf{r}_\alpha, \quad \mathbf{t}_{\theta,\alpha} = \mathbf{e}_y \times \mathbf{r}_\alpha = \mathbf{t}_{y,\alpha} \tag{3.58}$$

The usual angular velocities, ω_x, ω_y and ω_z, are inconvenient since the redundancy among them demands special precautions[59]. However, we can circumvent this problem completely by considering the momentum transformation.

From Eq. (3.57) it immediately follows that

$$\begin{aligned}
\mathscr{P}_F &= \sum_\alpha P_{\alpha F}; & F &= X, Y, Z \\
p_\varphi &= \sum_\alpha \mathbf{t}_{\varphi,\alpha} \cdot \mathbf{P}_\alpha, & p_\theta &= \sum_\alpha \mathbf{t}_{\theta,\alpha} \cdot \mathbf{P}_\alpha \\
p_k &= \sum_\alpha \mathbf{t}^0_{k,\alpha} \cdot \mathbf{P}_\alpha; & k &= 1, 2, \ldots 3N-5
\end{aligned} \tag{3.59}$$

But from the basic definition of angular momentum we also have

$$P_g = \sum_\alpha \mathbf{e}_g \cdot (\mathbf{r}_\alpha \times \mathbf{P}_\alpha) = \sum_\alpha \mathbf{t}_{g,\alpha} \cdot \mathbf{P}_\alpha; \; g = x, y, z \tag{3.60}$$

employing rotational t-vectors from Eq. (2.38) in rewriting. The basic condition on a set of generalized quasi momenta is that the momentum transformation must be invertible. In Eq. (3.59) we can therefore replace p_φ and p_θ by two of the angular momenta of Eq. (3.60) provided that the s-vectors of an inverse transformation can be determined. This is the case if we use P_x and P_y, giving

$$\mathbf{P}_\alpha = \sum_F \mathbf{s}_{F,\alpha}\, \mathscr{P}_F + \mathbf{s}_{x,\alpha} P_x + \mathbf{s}_{y,\alpha} P_y + \sum_k \mathbf{s}_{k,\alpha}\, p_k \tag{3.61}$$

The two sets of rotational s-vectors are easily expressed using the corresponding constraint vectors of the Eckart conditions as described in Sect. 2.2.3.2.2,

$$\mathbf{s}_{g,\alpha} = \eta\, m_\alpha\, \mathbf{e}_g \times \mathbf{a}_\alpha; \; g = x, y \tag{3.62}$$

where the same η applies to both axes,

$$\eta^{-1} = I^0 + \sum_\alpha m_\alpha\, a_{\alpha z}\, d_{\alpha z} \tag{3.63}$$

Introducing normal coordinates and l-vectors [Eq. (3.45)], we also have

$$\mathbf{s}_{k,\alpha} = \mathbf{s}^0_{k,\alpha} - \sum_{k'} Q_{k'} (\zeta^x_{k'k}\, \mathbf{s}_{x,\alpha} + \zeta^y_{k'k}\, \mathbf{s}_{y,\alpha}) \tag{3.64}$$

corresponding to Eq. (3.12) invoking Eq. (3.47).

Thus the remaining derivation can proceed exactly as for nonlinear molecules (Sect. 3.5), and the vibration-rotation part of the kinetic energy finally becomes

$$2\,T_{\text{vib-rot}} = \mu\,\{(P_x - p\!\!/_x)^2 + (P_y - p\!\!/_y)^2\} + \sum_k p_k^2 \tag{3.65}$$

with

$$\mu = I^0\,\eta^2 \tag{3.66}$$

and with vibrational angular momenta given by Eq. (3.38).

The angular momenta of Eq. (3.60) behave quite differently from those referring to a molecular coordinate system with three rotational degrees of freedom. In particular this relates to the commutation relations of the corresponding quantum mechanical operators. This is briefly discussed below in connection with the evaluation of the quantum kinetic energy.

By substituting $\mathbf{P}_\alpha$ [Eq. (3.61)] into the expressions of p_φ, p_θ and P_z [Eqs. (3.59)–3.60)] one obtains the relations

$$\begin{aligned} P_x &= -\csc\theta\, p_\varphi + \cot\theta\, p\!\!/_z \\ P_y &= p_\theta \\ P_z &= p\!\!/_z \end{aligned} \tag{3.67}$$

which apply to quantum momenta as well. Using Eq. (2.78) and $[p_q, f(q)] = -i\hbar\,\partial f(q)/\partial q;\ q = \varphi, \theta$ or Q_k, the commutation relations between the angular momenta are evaluated as

$$\begin{aligned} &[P_x, P_z] = [P_y, P_z] = 0 \\ &[P_x, P_y] = -i\hbar(\cot\theta\, P_x + P_z) \end{aligned} \tag{3.68}$$

and commutators involving directional cosines become

$$\begin{aligned} &[P_x, \Phi_{Fx}] = i\hbar\cot\theta\,\Phi_{Fy}, && [P_y, \Phi_{Fx}] = i\hbar\Phi_{Fz} \\ &[P_x, \Phi_{Fy}] = -i\hbar(\cot\theta\,\Phi_{Fx} + \Phi_{Fz}), && [P_y, \Phi_{Fy}] = 0, \qquad [P_z, \Phi_{Fg}] = 0 \\ &[P_x, \Phi_{Fz}] = i\hbar\Phi_{Fy}, && [P_y, \Phi_{Fz}] = -i\hbar\Phi_{Fx} \end{aligned} \tag{3.69}$$

Because of these anomalous commutation relations the transformation to space-fixed angular momentum components takes a special form. An equation similar to Eq. (2.80) yields

$$P_F = \sum_g \Phi_{Fg} P_g + \frac{i\hbar}{2}\cot\theta\,\Phi_{Fy} = \sum_g P_g \Phi_{Fg} - \frac{i\hbar}{2}\cot\theta\,\Phi_{Fy} \tag{3.70}$$

and the square of the total angular momentum is found accordingly,

$$P^2 = P_x^2 + \left(P_y - \frac{i\hbar}{2}\cot\theta\right)\left(P_y + \frac{i\hbar}{2}\cot\theta\right) + P_z^2$$
$$= P_x^2 + P_y^2 + P_z^2 - \frac{1}{4}\hbar^2 (1 + \csc^2\theta) \qquad (3.71)$$

Equation (3.69) is also important in connection with the U-function [Eq. (2.82)]. For linear molecules U is quite easily obtained, since all commutators involving vibrational momenta cancel. Thus from special properties of l-vectors and Coriolis coupling constants[57] it follows that

$$\sum_k \zeta_{kk}^g a_{k,\alpha}^{(gg)} = 0 \qquad (3.72)$$

from which it follows that

$$[\not{p}_g, \mu] = 0, \quad \sum_k [p_k, s_{k,\alpha}] = \mathbf{0}$$
$$\sum_k s_{k,\alpha} \cdot [p_k, s_{g,\alpha}] = 0 \qquad (3.73)$$

From Eq. (3.69) we then find that the only commutator of the type $[\pi_i, s_{i,\alpha F}]$ contributing to U is

$$[P_x, s_{x,\alpha F}] = i\hbar \cot\theta \, s_{y,\alpha F} \qquad (3.74)$$

and U is finally obtained as

$$U = -\frac{1}{8}\hbar^2 \mu (1 + \csc^2\theta) \qquad (3.75)$$

Adding this to the operator given by Eq. (3.65) and comparing it with Eq. (3.71), it can be seen that the kinetic energy operator can be written

$$T_{\text{vib-rot}} = \frac{1}{2}\mu\{P^2 - P_z^2 + \not{p}_x^2 + \not{p}_y^2 - P_x \not{p}_x - \not{p}_x P_x - P_y \not{p}_y - \not{p}_y P_y\}$$
$$+ \frac{1}{2}\sum_k p_k^2 \qquad (3.76)$$

or in a form corresponding to Watson's [59]

$$T_{\text{vib-rot}} = \frac{1}{2}\mu\left\{(P_x - \not{p}_x)^2 + \left(P_y - \not{p}_y - \frac{i\hbar}{2}\cot\theta\right)\left(P_y - \not{p}_y + \frac{i\hbar}{2}\cot\theta\right)\right\}$$
$$+ \frac{1}{2}\sum_k p_k^2 \qquad (3.77)$$

This rather extended discussion of the theory of a very special type of rigid molecule has been presented here, since it well illustrates the advantages obtained by a reformulation of the basic principles in terms of the momentum transformation. It applies to classical as well as quantum mechanical considerations. Furthermore, these examples, applying the method to wellknown molecular models, should make it easier to follow the derivations of the following section.

4 Hamiltonian of Nonrigid Molecules

For ordinary molecules the rigid rotor model works quite succussfully in explaining pure rotation spectra as well as the rotational fine structure obtained in other fields of spectroscopy. The vibrational perturbations appear mostly in the change of effective rotational constants with the vibrational state and in the centrifugal distortion effects. Much useful information can be found from these perturbation effects, however[5].

Nonrigid molecules have been studied similarly using semirigid rotor models, where internal motions with small amplitudes are ignored as in the rigid rotor[8,9]. Most efforts have been devoted to problems involving only one large amplitude motion. Thus methods of treating a single internal rotor, a single bending, inversion or a ring puckering are now discussed in several textbooks[66,67]. The results of such treatments are similar to the rigid rotor approach in that one obtains effective constants, including effective potential constants and reduced mass for the internal motion. But in general we still meet great difficulties when attempts are made to interpret these effective constants in relation to the fundamental structural and potential constants of the molecule.

As a step towards a more generally applicable theory this section will be concerned with generalizing the treatment of Sect. 3 to molecules where a single coordinate requires a special treatment. The result obtained can in principle be extended to cases with several special coordinates without difficulty.

First the Hamiltonian is derived and then an effective rotation – large amplitude motion operator is evaluated by a perturbation treatment similar to the treatment used for the rigid molecules.

4.1 Internal Coordinates

Recollecting the discussion of Eq. (2.7) we shall consider $3N-7$ rectilinear displacement coordinates, q_k; $k = 1, 2, \ldots 3N-7$, whereas the remaining internal coordinate, ρ, is specially treated by defining the reference configuration, given by the coordinates $a_{\alpha g}$, as well as the t-vector components, $t^0_{k,\alpha g}$, as appropriate functions of ρ. We write explicitly

$$r_{\alpha g} = a_{\alpha g}(\rho) + \sum_k t^0_{k,\alpha g}(\rho)\, q_k \tag{4.1}$$

The coordinates, $a_{\alpha g}(\rho)$, are the functions we would use in forming the Hamiltonian for the semirigid rotor model. Actually this well-known treatment is an important intermediate, and the general features must therefore be recapitulated.

4.2 Semirigid Rotor Approach

Neglecting all the small amplitude motions Eq. (2.11) yields

$$\dot{\mathbf{R}}_\alpha = \dot{\mathbf{R}} + \boldsymbol{\omega} \times \mathbf{a}_\alpha + \mathbf{t}^0_{\rho,\alpha}\dot{\rho} \tag{4.2}$$

where we have defined the vector function,

$$\mathbf{t}^0_{\rho,\alpha} = \frac{\partial \mathbf{a}_\alpha}{\partial \rho} \tag{4.3}$$

Neglecting translations the kinetic energy of the semirigid rotor may therefore be written

$$2\,T_{\text{S.R.}} = \sum_{gg'} I^0_{gg'}\,\omega_g\omega_{g'} + 2\sum_g I^0_{g\rho}\,\omega_g\,\dot{\rho} + I^0_{\rho\rho}\,\dot{\rho}^2 \tag{4.4}$$

where we have introduced a generalized "inertial" tensor of dimension 4 x 4, all elements of which may be functions of ρ. If the meaning of the label g is extended to include ρ as well, defining $\omega_\rho \equiv \dot{\rho}$, we can obtain the more general form,

$$\left.\begin{aligned} 2\,T_{\text{S.R.}} &= \sum_{gg'} I^0_{gg'}\omega_g\omega_{g'}, \\ I^0_{gg'} &= \sum_\alpha m_\alpha\,\mathbf{t}^0_{g,\alpha}\cdot\mathbf{t}^0_{g',\alpha}, \end{aligned}\right\}\quad g = x, y, z, \rho \tag{4.5}$$

with rotational t-vectors given by Eq. (3.8). We shall notice as we proceed below that the so-defined I^0-tensor takes over the important role which was previously played by the inertial tensor of the equilibrium configuration.

4.3 Constraints

Taking into consideration the small amplitude motions as well, we must consider the final decisive problem of how to define the relation between an arbitrary configuration and a reference. Besides the condition of a common center of mass we now need four constraint relations rather than three, since the reference configuration has a total of seven degrees of freedom. Hence, we must also establish a convention for the value of ρ to use in evaluating a_α-components. In general this ρ-value will deviate from the value, $\bar{\rho}$, applying to the actual configuration[17] (the set of r_α-vectors).

Our previous experiences with the rotational s-vectors suggest a relation between the set of s-vectors applying to ρ and the constraint vectors of a ρ-convention. Moreover, this implies a possibility for minimizing the couplings between the momentum P_ρ and the small amplitude momenta, p_k, arising from the elements of $\mathbf{G}$,

$$G_{\rho k} = \sum_{\alpha} m_\alpha^{-1} \mathbf{s}_{\rho,\alpha} \cdot \mathbf{s}_{k,\alpha} \; ; \; k = 1, 2 \ldots 3N-7 \tag{4.6}$$

In Eq. (2.35) we therefore consider particularly the relationship

$$\sum_{\alpha} \mathbf{t}_{\rho,\alpha} \cdot \mathbf{s}_{k,\alpha} = 0 \tag{4.7}$$

which using arguments similar to those justifying the Eckart conditions (Sect. 3.3) guide us to choose the constraint vectors

$$\mathbf{c}_{\rho,\alpha} = m_\alpha \, \mathbf{t}^0_{\rho,\alpha} \tag{4.8}$$

corresponding to Eq. (3.35). Since $\mathbf{t}^0_{\rho,\alpha}$ is independent of any q_k, we can evaluate an integral of the same form as Eq. (3.34), and hereby we obtain the constraint relation

$$\sum_{\alpha} m_\alpha \, \mathbf{t}^0_{\rho,\alpha} \cdot \mathbf{d}_\alpha = 0 \tag{4.9}$$

which actually is the Sayvetz condition[40].

As rotational constraints we retain the three Eckart conditions and the corresponding constraint vectors [Eq. (2.68) or (3.35)], here depending on ρ. These constraints have been discussed by Hougen[17] as well.

4.4 *s*-Vectors

Having established the basic constraints, the remaining derivations are of a purely algebraic nature.

As outlined in the previous sections, we can combine the treatment of the large amplitude motion and the rotations. Accordingly we start by extending the definitions of the *J*- and η-matrices. With the dimension 4 x 4 we now have

$$J_{gg'} = \sum_{\alpha} m_\alpha \, \mathbf{t}^0_{g,\alpha} \cdot \mathbf{t}_{g',\alpha} ; \; g = x, y, z, \rho \tag{4.10}$$

Retaining $\boldsymbol{\eta}$ as the designation for the inverse of $\mathbf{J}$ we thus obtain the *s*-vectors,

$$\mathbf{s}_{g,\alpha} = \sum_{g'} \eta_{gg'} \, m_\alpha \, \mathbf{t}^0_{g',\alpha} \tag{4.11}$$

for rotation and large amplitude motion, whereas the *s*-vectors of the small amplitude vibrations are found from zeroth order *s*-vectors and generalized Malhiot-Ferigle conditions exactly as in the case of rigid molecules (Sect. 3.1.1). The methods of Sect. 3.4 can also be transferred, and thus we introduce normal coordinates, $Q_1, Q_2, \ldots Q_{3N-7}$, with *s*- and *t*-vectors given by

$$\mathbf{s}_{k,\alpha} = m_\alpha^{1/2}\,\mathbf{l}_{k,\alpha} - \sum_{gk'} Q_{k'}\,\zeta_{k'k}^{g}\,\mathbf{s}_{g,\alpha}$$
$$\mathbf{t}_{k,\alpha} = m_\alpha^{-1/2}\,\mathbf{l}_{k,\alpha} \tag{4.12}$$

Here the rotational ζ-functions are given by Eq. (3.47), whereas the corresponding ζ-functions of ρ-coupling must be separately treated by considering the condition imposed on $\mathbf{s}_{k,\alpha}$ from Eq. (2.35),

$$\sum_\alpha \mathbf{s}_{k,\alpha} \cdot \mathbf{t}_{\rho,\alpha} = 0 \tag{4.13}$$

The general *t*-vectors of ρ are found by differentiating Eq. (4.1) giving

$$\mathbf{t}_{\rho,\alpha} = \frac{\partial \mathbf{r}_\alpha}{\partial \rho} = \mathbf{t}_{\rho,\alpha}^0 + m_\alpha^{-1/2} \sum_k Q_k \frac{\partial \mathbf{l}_{k,\alpha}}{\partial \rho} \tag{4.14}$$

and substituting this and Eq. (4.12) in the condition Eq. (4.13), we easily find

$$\zeta_{k'k}^{\rho} = \sum_\alpha \frac{\partial \mathbf{l}_{k',\alpha}}{\partial \rho} \cdot \mathbf{l}_{k,\alpha} = -\zeta_{kk'}^{\rho} \tag{4.15}$$

The last equality follows from Eq. (3.46). Also notice that Eq. (4.9) was used in deriving

$$\sum_\alpha m_\alpha^{1/2}\,\mathbf{l}_{k,\alpha} \cdot \mathbf{t}_{\rho,\alpha}^0 = \sum_\alpha m_\alpha\,\mathbf{t}_{\rho,\alpha}^0 \cdot \mathbf{t}_{k,\alpha} = 0 \tag{4.16}$$

4.5 Kinetic Energy

We generalize the procedure from Sect. 3.5 and use the momentum transformation as expressed by Eq. (3.49). However, when the range of g is extended to include ρ as well, another generalized quantity appears, namely

$$\not{p}_\rho = \sum_{kk'} Q_k\,\zeta_{kk'}^{\rho}\,\mathrm{p}_{k'} \tag{4.17}$$

This is a large amplitude momentum arising from the small amplitude motion in a way similar to that of the vibrational angular momenta [Eq. (3.38)]. The momentum vanishes in the reference configuration (all $Q_k = 0$) in accordance with our aim of removing zeroth order coupling effects.

Without further difficulties the kinetic energy can be given by an expression which is identical to Eq. (3.50) with a generalized μ-tensor of dimension 4 x 4,

$$\boldsymbol{\mu} = \boldsymbol{\eta}\,\mathbf{I}^0\,\boldsymbol{\eta} \tag{4.18}$$

The $\mathbf{I}^0$ appearing here is the generalized inertial tensor of the semirigid model [Eq. (4.5)]. The fact that Eq. (3.50) could be used in general was pointed out by Newton and Thomas[68] as early as 1948.

The problems to solve in dealing with specific molecular examples are therefore particularly connected with the ρ-dependence of the various quantities which formerly appeared as constants. We shall discuss some general properties of these functions.

4.6 Functions of ρ

Large amplitude motion requires special consideration when defining the coordinate functions $a_{\alpha g}(\rho)$; $g = x, y, z,\ \alpha = 1, 2, \ldots N$. Of course the relative positions of the atoms are given by the geometrical meaning of ρ, but the orientation of the molecular axes within a reference configuration can be defined in many ways. Our choice in this respect affects the elements of $\mathbf{I}^0$ and therefore the couplings between angular momenta and P_ρ. Actually this is a question of a convenient definition of the basic semirigid rotor model, and therefore we shall presently assume that our experience with this type of model can guide us in establishing the functional form of all $a_{\alpha g}(\rho)$.

Next we can consider the small amplitude motions which present a standard *GF*-eigenvalue problem, but with ρ as a free parameter. G^0-elements corresponding to a basic set of internal valence coordinates, R_t where $t = 1, 2, \ldots 3N{-}7$, are derived from s^0-vectors using Eq. (3.41), and thus they vary with ρ according to the variation of the first derivatives of Eq. (3.3). Also the force constants, $F_{tt'}$, may be functions of ρ and contribute to the general functional properties of L- and l-elements as well as of the eigenvalues, λ_k (Sect. 4.7).

These functions of ρ are special to the molecule in question. They may be expressed either analytically, as in the example in Sect. 5, or as a set of specific values derived numerically for appropriate values of ρ. Since numerical methods are most likely for larger molecules, we shall aim particularly at expressions where most partial derivates other than those of $\mathbf{a}_\alpha$ and $\mathbf{s}^0_{t,\alpha}$ have been eliminated. Numerically derived partial derivatives may introduce uncontrollable errors.

4.6.1 *J*- and μ-Functions

We shall first consider the J-matrix and in particular the elements introduced by including ρ-labelling. Comparing Eqs. (4.10) and (4.14) with the formulae for rigid molecules, Eqs. (3.14) and (3.51) in particular, we see that once more one may introduce a J'-matrix with elements linear in the normal coordinates of the small amplitude vibrations,

$$\left.\begin{aligned} J_{gg'} &= I^0_{gg'} + J'_{gg'} \\ J'_{gg'} &= \sum_k J^{(k)}_{gg'} Q_k \end{aligned}\right\} \quad g = x, y, x, \rho \qquad (4.19)$$

The rotational elements ($g = x, y$ or z) are given by Eq. (3.51) whereas we find

$$J_{\rho\rho}^{(k)} = -\sum_{\alpha} m_{\alpha}^{1/2}\, \mathbf{l}_{k,\alpha} \cdot \frac{\partial^2 \mathbf{a}_{\alpha}}{\partial \rho^2} \tag{4.20}$$

$$J_{g\rho}^{(k)} = \sum_{\alpha} m_{\alpha}^{1/2} (\mathbf{e}_g \times \mathbf{l}_{k,\alpha}) \cdot \frac{\partial \mathbf{a}_{\alpha}}{\partial \rho};\ g = x, y, z \tag{4.21}$$

In deriving these equations

$$\begin{aligned} 0 &= \frac{\partial}{\partial \rho}\left(\sum_{\alpha} m_{\alpha}^{1/2} \frac{\partial \mathbf{a}_{\alpha}}{\partial \rho} \cdot \mathbf{l}_{k,\alpha}\right) = \\ &\sum_{\alpha} m_{\alpha}^{1/2}\left(\frac{\partial \mathbf{a}_{\alpha}}{\partial \rho} \cdot \frac{\partial \mathbf{l}_{k,\alpha}}{\partial \rho} + \mathbf{l}_{k,\alpha} \cdot \frac{\partial^2 \mathbf{a}_{\alpha}}{\partial \rho^2}\right) \\ 0 &= \frac{\partial}{\partial \rho}\left(\sum_{\alpha} m_{\alpha}^{1/2} (\mathbf{e}_g \times \mathbf{a}_{\alpha}) \cdot \mathbf{l}_{k,\alpha}\right) = \\ &\sum_{\alpha} m_{\alpha}^{1/2}\left((\mathbf{e}_g \times \mathbf{a}_{\alpha}) \cdot \frac{\partial \mathbf{l}_{k,\alpha}}{\partial \rho} - (\mathbf{e}_g \times \mathbf{l}_{k,\alpha}) \cdot \frac{\partial \mathbf{a}_{\alpha}}{\partial \rho}\right) \end{aligned} \tag{4.22}$$

which follows from the Eckart- and Sayvetz-conditions.

The generalized J-matrix, and consequently μ as well, retain the property of being symmetric. μ is given by a generalized matrix equation similar to Eq. (3.17), or by

$$\begin{aligned} \boldsymbol{\mu} &= \boldsymbol{\mu}^0 + \boldsymbol{\mu}' + \frac{3}{4}\boldsymbol{\mu}'\, \mathbf{I}^0 \boldsymbol{\mu}' + \ldots \\ \boldsymbol{\mu}' &= -2\, \boldsymbol{\mu}^0\, \mathbf{J}'\, \boldsymbol{\mu}^0 \end{aligned} \tag{4.23}$$

Only for molecules of high symmetry is it possible to select the reference configurations such that diagonal I^0- and μ-matrices are obtained for any value of ρ. The matrix products of Eq. (4.23) may therefore be more involved than in the case of rigid molecules.

4.6.2 ζ-Functions

Functions of ρ corresponding to Coriolis-coupling constants are still determined by Eq. (3.47) which cannot be further simplified.

In the ζ-functions for ρ-coupling as given in Eq. (4.15) it would be convenient, however, if the l-vector derivatives were substituted by derivatives of more fundamental quantities. As such we shall consider the s^0-vectors, the G^0-elements or the basic force "constants" corresponding to the fundamental set of internal valence coordinates, R_t; $t = 1, 2, \ldots 3N-7$. These coordinates are related to the normal coordinates by the transformation

$$R_t = \sum_k L_{tk} Q_k \tag{4.24}$$

where the elements, L_{tk}, form the L-matrix of the GF-treatment. The s^0-vectors for the internal coordinates, R_t, are related to the l-vectors by

$$\mathbf{l}_{k,\alpha} = m_\alpha^{-1/2} \sum_t L_{kt}^{-1} \mathbf{s}_{t,\alpha}^0 \tag{4.25}$$

This equation is differentiated with respect to ρ and the result is substituted into Eq. (4.15), giving

$$\zeta_{k'k}^\rho = \sum_{\alpha t} m_\alpha^{-1/2} L_{k't}^{-1} \mathbf{l}_{k,\alpha} \cdot \frac{\partial \mathbf{s}_{t,\alpha}^0}{\partial \rho} + \sum_t \frac{\partial L_{k't}^{-1}}{\partial \rho} L_{tk} \tag{4.26}$$

The last term is evaluated considering the basic relations between G, F and L-matrices. After rather involved manipulations one obtains

$$\sum_t \frac{\partial L_{k't}^{-1}}{\partial \rho} L_{tk} = \frac{1}{\lambda_{k'} - \lambda_k} \sum_{tt'} \left[\lambda_k \frac{\partial G_{tt'}^0}{\partial \rho} L_{kt}^{-1} L_{k't'}^{-1} + \frac{\partial F_{tt'}}{\partial \rho} L_{tk} L_{t'k'} \right] \tag{4.27}$$

The ζ-expression resulting from Eqs (4.26) and (4.27) can be rewritten to a probably more useful form where the G^0-elements have been substituted by means of s^0-vectors,

$$\zeta_{k'k}^\rho = \frac{1}{\lambda_{k'} - \lambda_k} \left[\sum_{\alpha t} m_\alpha^{-1/2} (\lambda_{k'} L_{k't}^{-1} \mathbf{l}_{k,\alpha} + \lambda_k L_{kt}^{-1} \mathbf{l}_{k',\alpha}) \cdot \frac{\partial \mathbf{s}_{t,\alpha}^0}{\partial \rho} + \sum_{tt'} \frac{\partial F_{tt'}}{\partial \rho} L_{tk} L_{t'k'} \right] \tag{4.28}$$

4.7 Potential Energy

For rigid molecules it is convenient to express the potential energy by an expansion in powers of the internal displacement coordinates. A similar expansion can also be applied in the case of nonrigid molecules. However, the potential function is expanded only in the small amplitude coordinates, R_t; $t = 1, 2, \ldots 3N-7$ [17, 36],

$$V = V^0 + \sum_t \left. \frac{\partial V}{\partial R_t} \right|_0 R_t + \frac{1}{2} \sum_{tt'} \left. \frac{\partial^2 V}{\partial R_t \partial R_{t'}} \right|_0 R_t R_{t'} \ldots \tag{4.29}$$

where the derivatives are evaluated for all $R_t = 0$, i.e. for the ρ-dependent reference configurations. Hence, V^0 and all the derivatives are functions of ρ. V^0, in particular, represents the potential function of the semirigid model, e.g. the V_n-potential of a hindered internal rotor.

The second derivatives of the third term are the force "constants", $F_{tt'}$, which enter into the GF-treatment as discussed above. Consequently we can transform to a more convenient expression in the dimensionless normal coordinates,

$$q_k = Q_k\sqrt{2\pi c\omega_k/\hbar},\ 2\pi c\omega_k = (\lambda_k)^{1/2}$$

$$V = V^0 + \sum_k V^{(k)} q_k + \frac{1}{2}\sum_k hc\omega_k q_k^2 + \frac{1}{6}\sum_{kk'k''} hc\,\Phi_{kk'k''} q_k q_{k'} q_{k''} \ldots \qquad (4.30)$$

It has been conjectured that the second term, linear in the small amplitude coordinates, causes the most serious perturbations of the semirigid model. In a simple classical picture this term determines the actual ρ-dependence of the equilibrium bond lengths and angles which were assumed to be constant in the first approximation. This will be discussed in more detail below (Sect. 4.8.2.3). The effect has been considered by introducing the concept of relaxation in the semirigid model[21,34,35,69–71], i.e. by evaluating the I^0-matrix [Eq. (4.5)] from more elaborated coordinate functions, $a_{\alpha g}(\rho)$, such that not only a single valence coordinate varies with ρ.

Relaxation can be introduced in the present treatment as well. In principle it is always possible to eliminate the linear terms in V by defining the reference configurations as minimum energy configurations. This means that for all ρ it should hold that

$$\left.\frac{\partial V}{\partial R_t}\right|_0 = 0;\ t = 1, 2, \ldots 3N-7 \qquad (4.31)$$

However, from the point of view of simplifying the calculation of the kinetic energy, it is probably most advantageous to retain the linear terms and correct for their effects under the perturbation treatment. An example is given below (Sect. 5) in the discussion of the C_3-molecule.

4.8 Effective Semirigid Rotor Hamiltonian

Usually the vibration-rotation spectra of ordinary rigid molecules are analyzed in terms of effective rigid rotor Hamiltonians and vibrational energy expressions that result from a perturbation treatment. The Van Vleck transformation has been used for this purpose in various formulations. Thus the technique of successive contact transformations has been extensively utilized by Amat, Nielsen and Tarrago[4], whereas Jørgensen and Pedersen[72–74] recently suggested a formalism in terms of projection operators which offers advantages of generality combined with clarity. This technique has been applied in the present work.

In most cases the spectral analyses are based on Hamiltonians obtained by a quite simple perturbation treatment. Usually it is sufficient to carry the perturbations to an order where the rotational constants depend linearly on the vibrational quantum numbers, whereas the vibrational energies include quadratic terms. The relevant formulae have been discussed in the review by Mills[5]. Inadequacies of this approach are often encountered, but, in most cases, the deficiencies can be given two main reasons. The first is the presence of near degeneracies causing resonances of various types. The latter is that the molecule is nonrigid, i.e. that the special methods discussed in the present paper should be applied.

Extending the perturbation treatment to higher orders is possible, but extremely complicated. However, problems of the first kind may alternatively be solved by special methods comprising diagonalization of smaller matrix blocks.

Here nonrigidity will be considered under the perturbation treatment as well by excluding terms in the large amplitude coordinates and momenta from the zeroth order Hamiltonian. The resulting effective semirigid rotor Hamiltonians are therefore operators confined to the separate eigenspaces of a zeroth order Hamiltonian with terms of the small amplitude motion only [Eq. (4.36)].

At this introductory stage we can carry the comparison with the treatment of ordinary molecules further. In the first approximation these are described by the rigid rotor – harmonic oscillator model. In the next approximation an improvement is achieved by using effective operators with properties as described above. Similarly we may expect that the semirigid rotor – harmonic oscillator model for nonrigid molecules may be improved by introducing effective operators of the form,

$$\begin{aligned} H^{(v)} = hc \sum_k \omega_k \left(v_k + \frac{1}{2}\right) + hc \sum_{k \leqslant k'} x_{kk'} \left(v_k + \frac{1}{2}\right)\left(v_{k'} + \frac{1}{2}\right) \\ + hc \sum_{gg'} J_g B_{gg'}^{(v)} J_{g'} + \frac{1}{4} hc \sum_{gg'g''g'''} J_g J_{g'} \tau_{gg'g''g'''}^{(v)} J_{g''} J_{g'''} + V^{(v)} \end{aligned} \tag{4.32}$$

Here ω_k and $x_{kk'}$; $k = 1, 2, \ldots m$, are constants. $B_{gg'}^{(v)}$ and $\tau_{gg'g''g'''}^{(v)}$ correspond to the rotational and the centrifugal distortion constants, since $J_g = P_g/\hbar$ has been introduced for the angular and large amplitude momenta in units of $\hbar$ ($g = x, y, z, 1, 2, \ldots n$, $m + n = 3N{-}6$). It must be emphasized, however, that they are not constants, since they depend on the large amplitude coordinates, ρ_i; $i = 1. 2. \ldots n$. Furthermore the matrix of $B_{gg'}^{(v)}$-elements is in general non-diagonal. Despite this we shall call these quantities by there usual names, but use citation marks around the word "constant" to prevent confusion. In the present approach it is assumed that $B_{gg'}^{(v)}$ and the effective potential $V^{(v)}$ depend linearly on the vibrational quantum numbers, v_k.

The expression of $H^{(v)}$ appears to be quite general. It may apply to cases with several large amplitude coordinates as well as to the case of none, i.e. to rigid molecules (only isolated levels of the small amplitude vibrations will be considered here, however). This suggests that the Hamiltonian may be derived in a way which is closely similar to the well-estabilished methods for rigid molecules. In the following verification of this it is assumed again that only a single large amplitude coordinate needs to be considered ($n = 1$, $m = 3N{-}7$).

4.8.1 Orders of Magnitude

Usually we start out by separating the various terms in the Hamiltonian according to their orders of magnitude[5)]. In the present case it is impossible, however, to suggest a general ordering scheme, since the contributions from the large amplitude motion may change considerably from one molecule to the other.

The problems may be illustrated by considering the potential energy [Eq. (4.30)] separately. It is convenient to divide into two terms V_a and V_b, where only V_b depend on ρ,

$$V_a = \frac{hc}{2} \sum_k \omega_k q_k^2 + \frac{hc}{6} \sum_{kk'k''} \Phi^0_{kk'k''} q_k q_{k'} q_{k''} + V_a^{(4)} \ldots \tag{4.33}$$

$$\begin{aligned} V_b = {} & V^0(\rho) + \sum_k V^{(k)}(\rho) q_k + \sum_k \frac{h}{8\pi^2 c\omega_k} (\lambda_k(\rho) - \lambda_k^0) q_k^2 \\ & + \frac{hc}{6} \sum_{kk'k''} (\Phi_{kk'k''}(\rho) - \Phi^0_{kk'k''}) q_k q_{k'} q_{k''} + V_b^{(4)}(\rho) \cdots \end{aligned} \tag{4.34}$$

Here $\omega_k = (2\pi c)^{-1}\sqrt{\lambda_k^0}$, λ_k^0, $\Phi^0_{kk'k''}$ and similar quantities of $V_a^{(4)}$ and higher order terms are constants chosen as a sort of mean value of the ρ-dependent quantities λ_k, $\Phi_{kk'k''}$ and so on, in order to minimize the perturbations from the corresponding terms in V_b. Notice, that now we use the constant λ_k^0 in the transformation to dimensionless normal coordinates,

$$q_k = Q_k \hbar^{-1/2} (\lambda_k^0)^{1/4} = Q_k \sqrt{2\pi c\omega_k/\hbar} \tag{4.35}$$

The properties of V_a may be expected to be very similar to what we observe for rigid molecules, and consequently we can use the general ordering scheme according to the powers of q_k for V_a. The first term contributes to H_0, the next to H_1 and so on.

When considering V_b it also seems reasonable to assume that if V^0 contributes in some order of magnitude, n, then another term, $V_b^{(m)}$, involving products of q_k 's to a total power of m, contributes in the order $m + n$. Hence the problem is to choose an appropriate n for V^0. On the other hand, V^0 should contribute to the same order of magnitude as the leading term in the expansion of the large amplitude kinetic energy, $\frac{1}{2} \mu^0_{\rho\rho} P_\rho^2$. For convenience and to emphasize the similarity with the treatment of ordinary molecules we shall therefore choose the same order as for the leading terms of the rotational energy, i.e. the order two. However, another ordering scheme specially adapted to the case of water was suggested by Hoy and Bunker[37].

The Hamiltonian is now split up as follows,

$$\begin{aligned} H_0 &= \frac{hc}{2} \sum_k \omega_k (p_k^2 + q_k^2) \\ H_1 &= V_a^{(3)} \\ H_2 &= V_a^{(4)} + \frac{1}{2} \sum_{gg'} (P_g - p_g) \mu^0_{gg'} (P_{g'} - p_{g'}) + V^0 \\ H_3 &= V_a^{(5)} + \frac{1}{2} \sum_{gg'k} (P_g - p_g) \mu^{(k)}_{gg'} q_k (P_{g'} - p_{g'}) + V_b^{(1)} \\ H_4 &= V_a^{(6)} + \frac{1}{4} \sum_{gg'kk'} (P_g - p_g) \mu^{(kk')}_{gg'} q_k q_{k'} (P_{g'} - p_{g'}) + V_b^{(2)} \end{aligned} \tag{4.36}$$

4.8.2 Van Vleck Transformation

The zeroth order Hamiltonian is a sum of one-dimensional harmonic oscillator operators. Eigenvalues and eigenfunctions of H_0 are designated according to the equation,

$$H_0 \mid V\rangle = \mid V\rangle \sum_k hc\omega_k \left(v_k + \frac{1}{2}\right) = \mid V\rangle\ E_v$$
$$\mid V\rangle \quad = \mid v_1, v_2, \ldots v_k, \ldots v_{3N-7}\rangle \tag{4.37}$$

where $v_k = 0, 1, 2, \ldots$ is a harmonic oscillator quantum number.

Following the technique of Jørgensen and Pedersen[72] we introduce projection operators on the complete Hilbert space of the Hamiltonian, the space spanned by product functions, $\{\mid V\rangle \cdot \mid L\rangle\}$, where $\mid L\rangle = \mid J, K, M, N_\rho\rangle$ is a basis function for the pure rotation – large amplitude motion problem. Aiming at an effective operator on a given eigenspace, Ω_v, of H_0 with the eigenvalue E_v, we define the projectors P_v and Q_v by

$$H_0 P_v = E_v P_v, \ Q_v = 1 - P_v \tag{4.38}$$

In terms of the eigenvectors of H_0 we can write,

$$P_v = \sum_L \mid V, L\rangle\langle V, L \mid = \sum_L \mid L\rangle\langle L \mid \prod_k \mid v_k\rangle\langle v_k \mid \tag{4.39}$$

It is convenient to define an operator, a_v, by

$$a_v = E_v - H_0 \tag{4.40}$$

and furthermore we introduce

$$\frac{Q_v}{a} = Q_v\, a_v^{-1}\, Q_v \tag{4.41}$$

With these symbols the most important terms of the effective operator can be expressed by

$$H_0^{(v)} = E_v P_v$$
$$H_2^{(v)} = P_v \left\{ H_1 \frac{Q_v}{a} H_1 + H_2 \right\} P_v \tag{4.42}$$
$$H_4^{(v)} = P_v \left\{ H_4 + H_2 \frac{Q_v}{a} H_2 + H_1 \frac{Q_v}{a} H_3 + H_3 \frac{Q_v}{a} H_1 \right\} P_v$$

Compared to the general expression in Ref.[72] Eq. (4.42) has been considerably reduced. First it has been taken into account that $H_n^{(v)}$ vanishes for odd n, since H_1 and H_3 are odd operators, whereas H_2 and H_4 are even with respect to the inversion of a vibrational coordinate, q_k. Secondly we shall neglect corrections to the small amplitude vibrational energies of higher order than two. This is the reason for omitting several terms in the original $H_4^{(v)}$. Some of the terms arising from this reduced equation still contribute outside the frame of Eq. (4.32). This is the case for some of the terms involving the vibrational momenta, $\not{p}_g$, and for the higher order anharmonicities $V_a^{(5)}$ and $V_a^{(6)}$. These corrections will be neglected as well.

Finally the present treatment is restricted to cases where essential vibrational degeneracies between the level under consideration, E_v, and other levels are absent.

4.8.2.1 Second Order Corrections

From the second order term, $H_2^{(v)}$, we obtain the $x_{kk'}$-quantities and the major contributions to $B_{gg'}^{(v)}$ and $V^{(v)}$ which are independent of the vibrational state. Readers not interested in details of the evaluation may proceed to Eqs. (4.50–51) and the summary after Eq. (4.57).

From $P_v H_2 P_v$ we get, using $u_k = v_k + \frac{1}{2}$,

$$P_v V_a^{(4)} P_v = P_v \sum_{kk'k''k'''} \frac{hc}{24} \Phi^0_{kk'k''k'''} \, q_k q_{k'} q_{k''} q_{k'''} P_v =$$

$$hc \left[\sum_k \Phi^0_{kkkk} \frac{1}{16} \left(u_k^2 + \frac{1}{4} \right) + \sum_{k<l} \Phi^0_{kkll} \frac{1}{4} u_k u_l \right] P_v \tag{4.43}$$

$$P_v \left[\frac{1}{2} \sum_{gg'} P_g \mu^0_{gg'} P_{g'} + V^0 \right] P_v = hc \left[\sum_{gg'} J_g B^0_{gg'} J_{g'} + V^0 \right] P_v \tag{4.44}$$

and

$$P_v \frac{1}{2} \sum_{gg'} \not{p}_g \mu^0_{gg'} \not{p}_{g'} P_v =$$

$$P_v \frac{1}{2} \hbar^2 \sum_{gg'} \mu^0_{gg'} \sum_{kl} \zeta^g_{kl} \zeta^{g'}_{kl} \left(\frac{\omega_l}{\omega_k} q_k^2 p_l^2 - q_k p_l q_l p_k \right) P_v =$$

$$hc \sum_{gg'} B^0_{gg'} \sum_{k<l} \zeta^g_{kl} \zeta^{g'}_{kl} \left[\left(\frac{\omega_l}{\omega_k} + \frac{\omega_k}{\omega_l} \right) u_k u_l - \frac{1}{2} \right] P_v \tag{4.45}$$

The remaining second order perturbations involving H_1 are evaluated by splitting $V_a^{(3)}$ into three terms,

$$V_a^{(3)} = hc \left[\frac{1}{6} \sum_k \Phi^0_{kkk} q_k^3 + \frac{1}{2} \sum_{k \neq l} \Phi^0_{kll} q_k q_l^2 + \sum_{k<l<m} \Phi^0_{klm} q_k q_l q_m \right] \tag{4.46}$$

We then obtain

$$P_\nu V_a^{(3)} \frac{Q_\nu}{a} V_a^{(3)} P_\nu = \Bigg\{ -\sum_k (\Phi^0_{kkk})^2 \frac{5\,hc}{48\,\omega_k} \left(u_k^2 + \frac{1}{4}\right)$$

$$-\sum_{k \neq l} \Phi^0_{kkk}\Phi^0_{kll} \frac{hc}{4\,\omega_k} u_k u_l - \sum_{k \neq l \neq m} \Phi^0_{kll}\,\Phi^0_{kmm} \frac{hc}{8\,\omega_k} u_l u_m$$

$$-\sum_{k \neq l} (\Phi^0_{kll})^2 \left[\frac{hc}{32}\left(\frac{4}{\omega_k} + \frac{1}{2\,\omega_l + \omega_k} - \frac{1}{2\,\omega_l - \omega_k}\right) u_l^2\right.$$

$$\left. + \frac{hc}{8}\left(\frac{1}{2\,\omega_l + \omega_k} + \frac{1}{2\,\omega_l - \omega_k}\right) u_k u_l + \frac{3\,hc}{128}\left(\frac{1}{2\,\omega_l + \omega_k} - \frac{1}{2\,\omega_l - \omega_k}\right)\right]$$

$$+ \sum_{k<l<m} (\Phi^0_{klm})^2 \frac{hc}{8} \left[\frac{-u_k u_l - u_l u_m - u_m u_k - \frac{1}{4}}{\omega_k + \omega_l + \omega_m} + \frac{u_k u_l - u_l u_m + u_m u_k - \frac{1}{4}}{\omega_k - \omega_l - \omega_m}\right.$$

$$\left. + \frac{u_l u_m - u_m u_k + u_k u_l - \frac{1}{4}}{\omega_l - \omega_m - \omega_k} + \frac{u_m u_k - u_k u_l + u_l u_m - \frac{1}{4}}{\omega_m - \omega_k - \omega_l}\right]\Bigg\} P_\nu \qquad (4.47)$$

This can be reduced by appropriately changing the indices and releasing the summation restrictions,

$$P_\nu V_a^{(3)} \frac{Q_\nu}{a} V_a^{(3)} P_\nu = -hc \Bigg\{ \sum_{kl} u_k^2 (\Phi^0_{kkl})^2 \frac{1}{32} \left(\frac{4}{\omega_l} + \frac{1}{2\,\omega_k + \omega_l} - \frac{1}{2\,\omega_k - \omega_l}\right)$$

$$+ \sum_{k<l} u_k u_l \sum_m \left(\Phi^0_{kkm}\,\Phi^0_{llm} \frac{1}{4\,\omega_m} + (\Phi^0_{klm})^2 \frac{1}{8} R_{klm}\right)$$

$$+ \sum_k (\Phi^0_{kkk})^2 \frac{1}{48\,\omega_k} - \sum_{klm} (\Phi^0_{klm})^2 \frac{1}{192} S_{klm} \qquad (4.48)$$

$$+ \sum_{kl} (\Phi^0_{kkl})^2 \frac{1}{128} \left(\frac{4}{\omega_l} + \frac{1}{2\,\omega_k + \omega_l} - \frac{1}{2\,\omega_k - \omega_l}\right)\Bigg\} P_\nu$$

where

$$R_{klm} = \frac{1}{\omega_k + \omega_l + \omega_m} + \frac{1}{\omega_k - \omega_l + \omega_m} + \frac{1}{\omega_l - \omega_k + \omega_m} + \frac{1}{\omega_m - \omega_k - \omega_l}$$

$$= 4\,\omega_m (\omega_m^2 - \omega_k^2 - \omega_l^2) / \Delta_{klm}$$

$$S_{klm} = \frac{1}{\omega_k + \omega_l + \omega_m} + \frac{1}{\omega_k - \omega_l - \omega_m} + \frac{1}{\omega_l - \omega_m - \omega_k} + \frac{1}{\omega_m - \omega_k - \omega_l} \qquad (4.49)$$

$$= -4\,\omega_k \omega_l \omega_m / \Delta_{klm}$$

$$\Delta_{klm} = (\omega_k + \omega_i + \omega_m)(\omega_k - \omega_l - \omega_m)(\omega_l - \omega_m - \omega_k)(\omega_m - \omega_k - \omega_l)$$

Before the terms are collected it should be noticed that the contribution in Eq. (4.45) is a function of the large amplitude coordinate, since $B^0_{gg'}$ as well as $\zeta^g_{kk'}$ depends on ρ. This dependence can be neglected, however, to the same degree of approximation as we are neglecting the differences, $\Phi - \Phi^0$, of V_b, [Eq. (4.34)]. It is therefore assumed that an appropriate mean value is used in x_{kl} below, and that the part which is independent of the vibrational state can be considered as an inessential constant contribution to the energy in the same way as similar terms of Eqs. (4.43) and (4.48).

Hence the anharmonic corrections to the vibrational energies are obtained from the second term of Eq. (4.32) using $x_{kk'}$-constants which for $k' = k$ are given by,

$$x_{kk} = \frac{1}{16}\,\Phi^0_{kkkk} - \frac{1}{32}\sum_l (\Phi^0_{kkl})^2 \left(\frac{4}{\omega_l} + \frac{1}{2\,\omega_k + \omega_l} - \frac{1}{2\,\omega_k - \omega_l}\right) \tag{4.50}$$

whereas for $k' = l > k$ we have,

$$x_{kl} = \frac{1}{4}\,\Phi^0_{kkll} - \sum_m \left(\Phi^0_{kkm}\,\Phi^0_{llm}\,\frac{1}{4\,\omega_m} + (\Phi^0_{klm})^2 R_{klm}/8\right) + \sum_{gg'} B^0_{gg'}\,\zeta^g_{kl}\,\zeta^{g'}_{kl}\left(\frac{\omega_l}{\omega_k} + \frac{\omega_k}{\omega_l}\right) \tag{4.51}$$

The only difference from the usual expressions[5)] appears in the last term of Eq. (4.51) where we must consider a possibly nondiagonal matrix of $B^0_{gg'}$-elements with the dimension 4 x 4 ($g = x, y, z, \rho$).

4.8.2.2 Fourth Order Corrections

From $H_4^{(v)}$ we obtain vibrational corrections to the "rotational constants" and the potential. First $P_v H_4 P_v$ gives

$$P_v H_4 P_v = \left[\frac{hc}{2}\sum_{kgg'} J_g B^{(kk)}_{gg'} u_k J_{g'} + \sum_k \frac{h(\lambda_k - \lambda^0_k)}{8\,\pi^2 c\,\omega_k}\,u_k\right] P_v, \tag{4.52}$$

where Eq. (4.23) must be applied in evaluating $B^{(kk)}_{gg'}$,

$$B^{(kk')}_{gg'} = \frac{3\,h}{4\,\pi^2 c}\,(\boldsymbol{\mu}^0\,\mathbf{J}^{(k)}\boldsymbol{\mu}^0\,\mathbf{J}^{(k')}\boldsymbol{\mu}^0)_{gg'} \tag{4.53}$$

If $\boldsymbol{\mu}^0$ is diagonal this reduces to

$$B^{(kk)}_{gg'} = \frac{3\,B^0_g B^0_{g'}}{\omega_k}\sum_{g''}\frac{a_k^{(gg'')}\,a_k^{(g'g'')}}{I^0_{g''}}, \quad a_k^{(gg')} = \frac{\partial I^*_{gg'}}{\partial Q_k} \tag{4.54}$$

The term in $B_{gg'}^{(kk)}$ of Eq. (4.52) corresponds to the usual "harmonic" contribution to the effective rotational constants.

From $P_v H_2 \frac{Q_v}{a} H_2 P_v$ we similarly get the "Coriolis coupling" contribution, using the term, Ω, of H_2,

$$\begin{aligned}\Omega &= -\frac{1}{2} \sum_{gg'} (P_g \mu_{gg'}^0 \, p_{g'} + p_{g'} \, \mu_{gg'}^0 P_g) \\ &= -\frac{\hbar}{2} \sum_{gg'} \sum_{k<l} (P_g \mu_{gg'}^0 \, \zeta_{kl}^{g'} + \mu_{gg'}^0 \, \zeta_{kl}^{g'} P_g) \left(q_k p_l \sqrt{\frac{\omega_l}{\omega_k}} - q_l p_k \sqrt{\frac{\omega_k}{\omega_l}} \right)\end{aligned} \tag{4.55}$$

We thus obtain,

$$\begin{aligned}P_v \Omega \frac{Q_v}{a} \Omega P_v = \frac{h}{8\pi^2 c} \sum_{gg'g''g'''kl} \frac{3\,\omega_k^2 + \omega_l^2}{\omega_k(\omega_k^2 - \omega_l^2)} & \Big(P_g \mu_{gg''}^0 \, \mu_{g'g'''}^0 \, \zeta_{kl}^{g''} \zeta_{kl}^{g'''} P_{g'} \\ & + \frac{1}{2} [P_g, \mu_{gg''}^0 \, \zeta_{kl}^{g''} [P_{g'}, \mu_{g'g'''}^0 \, \zeta_{kl}^{g'''}]] \\ & - \frac{1}{4} [P_g, \mu_{gg''}^0 \, \zeta_{kl}^{g''}] [P_{g'}, \mu_{g'g'''}^0 \, \zeta_{kl}^{g'''}] \Big) \; u_k P_v\end{aligned} \tag{4.56}$$

The commutators of this equation vanish in cases where P_g refers to an angular momentum. But, if P_g is a momentum of an internal large amplitude motion, the commutation terms may contribute to the potential $V^{(v)}$ as well as to the vibrational energies. The order of magnitude of these corrections is expected to be very small, however, and they will be neglected in the following.

The "anharmonic" contributions to the effective "rotational constants" and potential arise from the last two terms of $H_4^{(v)}$ in Eq. (4.42) giving,

$$\begin{aligned}& P_v \left(H_1 \frac{Q_v}{a} H_3 + H_3 \frac{Q_v}{a} H_1 \right) P_v = \\ & P_v \left(-\frac{1}{2} \sum_{gg'kl} P_g \mu_{gg'}^{(l)} \, \Phi_{kkl}^0 \frac{u_k}{2\,\omega_l} P_{g'} - \sum_{kl} V^{(l)} \Phi_{kkl}^0 \frac{u_k}{2\,\omega_l} \right)\end{aligned} \tag{4.57}$$

Summarizing these results we conclude that the "rotational constants" and the potential applying to Eq. (4.32) can be written in the form,

$$\begin{aligned}B_{gg'}^{(v)} &= B_{gg'}^0 - \sum_k \alpha_k^{(gg')} \left(v_k + \frac{1}{2} \right) \\ V^{(v)} &= V^0 - \sum_k \alpha_k^{(V)} \left(v_k + \frac{1}{2} \right)\end{aligned} \tag{4.58}$$

where

$$-\alpha_k^{(gg')} = \frac{1}{2} B_{gg'}^{(kk)} + 2 \sum_{g''g'''} B_{gg''}^0 B_{g'g'''}^0 \sum_l \left[\pi \left(\frac{c}{h}\right)^{1/2} a_l^{(g''g''')} \frac{\Phi_{kkl}^0}{\omega_l^{3/2}} + \zeta_{kl}^{g''} \zeta_{kl}^{g'''} \frac{3\,\omega_k^2 + \omega_l^2}{\omega_k(\omega_k^2 - \omega_l^2)} \right] \tag{4.59}$$

A corresponding α-quantity has been introduced for the large amplitude potential,

$$-\alpha_k^{(V)} = \frac{h}{8\,\pi^2 c\omega_k} (\lambda_k - \lambda_k^0) - \sum_l V^{(l)} \Phi_{kkl}^0 \frac{1}{2\,\omega_l} \tag{4.60}$$

It was noticed in the preceding paragraph that the $x_{kk'}$-formulae [Eq. (4.50) and (4.51)] apply to rigid molecules as well. Similarly Eq. (4.59) reduces to the usual α-expression if a diagonal $B_{gg'}^0$-matrix can be assumed, implying Eq. (4.54).

4.8.2.3 Higher Order Corrections

Two terms from H_3 may in some cases appear to be underestimated as to their order of magnitude. These are

$$H_3' = \frac{1}{2} \sum_{gg'k} P_g \mu_{gg'}^{(k)} q_k P_{g'} + \sum_k V^{(k)} q_k \tag{4.61}$$

If H_3' is raised one step in the ordering scheme, we must consider an additional fourth order correction,

$$P_v H_3' \frac{Q_v}{a} H_3' P_v = \left[-\frac{1}{8} \sum_{kgg'g''g'''} P_g \mu_{gg'}^{(k)} P_{g'} P_{g''} \mu_{g''g'''}^{(k)} P_{g'''} \frac{1}{hc\omega_k} - \frac{1}{2} \sum_{kgg'} P_g \mu_{gg'}^{(k)} V^{(k)} P_{g'} \frac{1}{hc\,\omega_k} - \frac{1}{2} \sum_k (V^{(k)})^2 \frac{1}{hc\,\omega_k} \right] P_v \tag{4.62}$$

Otherwise this correction appears in the sixth order. In the first term we recognize the centrifugal distortion correction which is important in any case. The second term gives a correction to $B_{gg'}^{(v)}$ and the third term contributes to the effective potential $V^{(v)}$. All of these are independent of the vibrational state.

The centrifugal distortion term, H_{CD}, can be rewritten as follows,

$$H_{\text{CD}} = \frac{1}{4} \sum_{gg'g''g'''} P_g P_{g'} \tau_{gg'g''g'''} P_{g''} P_{g'''} - \frac{1}{2} \sum_{gg'} P_g \xi_{gg'} P_{g'} \tag{4.63}$$

where

$$\tau_{gg'g''g'''} = -\sum_k \mu_{gg'}^{(k)} \mu_{g''g'''}^{(k)} (2\,hc\,\omega_k)^{-1} \tag{4.64}$$

$$\xi_{gg'} = \sum_{kg''g'''} \mu_{gg''}^{(k)} [P_{g''}, [P_{g'''}, \mu_{g'g'''}^{(k)}]] (4\,hc\omega_k)^{-1} \tag{4.65}$$

The question as to the importance of the commutators in Eq. (4.65) is similar to the problem discussed in relation to Eq. (4.56). For nonlinear molecules it should be safe to ignore the ρ-dependence of $\mu_{gg'}^{(k)}$ and thus to consider $\tau_{gg'g''g'''}$ as constant and to neglect $\xi_{gg'}$-terms. For linear molecules more care should be taken, since the μ_{zg}-elements depend inversely on a bending angle, giving singularities for the linear configuration. These singularities require special attention.

The remaining two terms of Eq. (4.62) are interesting as well, since they form the connection between molecular models incorporating relaxation, as discussed in Sect. 4.7, and the simpler models.

The physical meaning of the two terms may be visualized by arguments of classical physics in a way which is very similar to the explanation given for centrifugal distortion in rigid molecules[67]. Thus, if we allow the molecule to follow a minimum energy path in course of the large amplitude motion and rotation, i.e. if we allow the small amplitude coordinates to relax, then the changes in these coordinates are determined by the condition of minimum energy.

With an approximate Hamiltonian in which the small amplitude kinetic energy is ignored,

$$H' = V^0 + \sum_k V^{(k)} q_k + \frac{1}{2} hc \sum_k \omega_k q_k^2 + \frac{1}{2} \sum_{gg'} (\mu_{gg'}^0 + \sum_k \mu_{gg'}^{(k)} q_k) P_g P_{g'} \tag{4.66}$$

we have the condition,

$$\frac{\partial H'}{\partial q_k} = V^{(k)} + hc\,\omega_k q_k + \frac{1}{2} \sum_{gg'} \mu_{gg'}^{(k)} P_g P_{g'} = 0 \tag{4.67}$$

This equation can be set up and solved for any q_k, and if the solutions are substituted into Eq. (4.66) we obtain,

$$H' = V^0 - \sum_k (V^{(k)})^2 \frac{1}{2\,hc\,\omega_k} + \frac{1}{2} \sum_{gg'} \left(\mu_{gg'}^0 - \sum_k \mu_{gg'}^{(k)} V^{(k)} \frac{1}{hc\,\omega_k} \right) P_g P_{g'} - \frac{1}{8} \sum_{kgg'g''g'''} \mu_{gg'}^{(k)} \mu_{g''g'''}^{(k)} \frac{1}{hc\,\omega_k} P_g P_{g'} P_{g''} P_{g'''} \tag{4.68}$$

Hence the approximate rotation – large amplitude motion Hamiltonian takes the form of the second order operator, [Eq. (4.44)] with corrections exactly as those obtained from H_3' in Eq. (4.62). If we suppose that these perturbations are essential, we must replace $B_{gg'}^0$ and V^0 in Eq. (4.58) by

$$\begin{aligned} B_{gg'}^{e} &= B_{gg'}^0 - \sum_k B_{gg'}^{(k)} V^{(k)} \frac{1}{hc\,\omega_k} \\ V^{e} &= V^0 - \sum_k (V^{(k)})^2 \frac{1}{2\,hc\,\omega_k} \end{aligned} \tag{4.69}$$

where superscript e, for equilibrium, indicates that these quantities refer to configurations where the small amplitude coordinates have relaxed to their ρ-depending equilibrium values.

The Hamiltonian of Eq. (4.68) corresponds directly to the Hamiltonian for a semirigid model with relaxation[21, 34, 35, 69–71]. Such models have been introduced in often successful attempts to obtain better agreement between experiment and theory. But, in addition, it was hoped that a more reliable picture of the large amplitude potential function would result. This seems doubtful, however, considering that corrections of the fourth order are neglected. Even though all $V^{(k)}$-functions vanish, if reference structures with the correct relaxations are used, the terms in $\lambda_k - \lambda_k^0$ still survives in $\alpha_k^{(V)}$, [Eq. (4.60)], and all $\alpha_k^{(gg')}$ are unchanged, [Eq. (4.59)].

The relaxation in semirigid models is usually formulated in terms of parameters which are subsequently examined by experiment. The considerations above also imply that we should not overestimate the physical significance of such parameters. They only serve to simulate the ρ-dependence of the total contributions from the relaxation as well as the perturbations accounted for in the α-terms of Eq. (4.58).

4.9 Summary and Discussion

This completes the theoretical part of the present paper. It may be useful to recapitulate the principal aims and the main results.

It has been attempted to extend the standard theory for rigid molecules to molecules with large amplitude internal motions in a formulation which makes it possible to focus on general as well as special properties.

The s- and t-vectors of the momentum and velocity transformations [Eqs. (2.33), (2.44), (3.48), (3.49) and 3.61)], were found to be very effective for this purpose as demonstrated by their applicability in deriving the kinetic energy of ordinary rigid molecules (Sect. 3), as well as in discussing the more complex problems of the nonrigid molecules (Sect. 4). It is a special accomplishment of the present work that the principles of forming s-vectors corresponding to rotations and large amplitude internal motions have been established, (Sects. 2.2.3.2 and 4.3) in a way which clearly shows their connection with the basic constraints on the small amplitude vibrations relative to the reference structures. Also notice that the factorized expression for the μ-tensor [Eq. (2.55)], which has been observed in several special cases[47, 57, 58], appears as a general consequence of the method of constructing these s-vectors.

The Eckart- and Sayvetz-conditions constitute a set of conventions for the reference structures which are particularly useful, since they allow us to use rectilinear coordinates for the small amplitude motions (Sect. 3.3). However, the introduction of reference structures, depending on the large amplitude coordinates only, leaves us with the question of how the molecular axes should be oriented within an arbitrary set of atomic reference positions. This question was only briefly commented on in Sect. 4.6, since it is special to the molecule under consideration. Some examples may illustrate types of solutions.

For molecules with internal rotation[8,75] it is often convenient to use a frame-fixed set of axes which, in case of a symmetrical internal rotor can be chosen as a principal axis system for the reference structures (P.A.M.). In this case all off-diagonal elements vanish in the rotational part of $\boldsymbol{\mu}^0$, i.e. $\mu^0_{gg'} = \mu^0_g \delta_{gg'}$, $g = x, y, z$, whereas there may be nonvanishing coupling elements, $\mu^0_{g\rho}$, between internal and overall rotation. For molecules with a heavy symmetrical internal rotor on a light asymmetric frame, e.g. methanol, the Internal Axis Method has been proposed. Here the axes are chosen in the opposite way so that the coupling terms, $\mu^0_{g\rho}$, vanish, whereas the rotational part of $\boldsymbol{\mu}^0$ is nondiagonal.

Similar alternatives may be suggested for linear molecules with a large amplitude bending vibration. Thus Hougen et al.[17] and Sarka[47] have developed the semirigid reference models for three and four atomic linear molecules with an axis convention aiming at the elimination of $\mu^0_{g\rho}$-elements, i.e. corresponding to I.A.M. of internal rotation. Here the only nonvanishing element of $\boldsymbol{\mu}^0$ is μ^0_{yz}, if the x-axis is defined as perpendicular to the plane of the bent reference structures. A solution corresponding to P.A.M. will cause μ^0_{yz} to vanish but introduce a coupling term $\mu^0_{x\rho}$ [16, 44, 46]. None of these methods have particular advantages when considering a complete treatment including the evaluation of the effective operator [Eq. (4.32)], since contributions to $B^{(v)}_{yz}$ as well as $B^{(v)}_{x\rho}$ may arise from the perturbations in both cases.

In between there exists a third possibility for axis orientation in such molecules. This is to fix the z-axis to the axis of a linear configuration ($\rho = 0$), which is defined in relation to the instantaneous reference by Eckart-conditions. This method offers the special advantage that the coordinate functions, $a_{\alpha g}(\rho)$, can be expressed in quite simple closed forms so that derivatives of the coordinates as well as the vibrational s^0-vectors can be easily derived[76]. This more than counterbalances the fact that both μ^0_{yz} and $\mu^0_{x\rho}$ may be nonvanishing, since the appearance of the coupling terms $B^{(v)}_{yz}$ and $B^{(v)}_{x\rho}$ in the effective bending-rotation operator should not cause serious troubles. They may either be treated as perturbations or, if one of the couplings is dominating, we may eliminate that term by a final rotation of axes following the prescriptions of Pickett[77]. Picket also suggests the Eckart system as an intermediate for the final "Rational Axis System".

In relation to the Van Vleck transformation (Sect. 4.8) we recapitulate that most of the formulae applying to rigid molecules could be generalized with only small adjustments [Eqs. (4.32), (4.50), (4.51), 4.58)–(4.60), (4.63)–(4.65) and (4.69)]. This indicates that the treatment without particular complications may be extended to cover a case where the small amplitude vibrational level is degenerate. This, however, is an object for future developments.

For nonrigid molecules there are at present only few reported data on excited states of the small amplitude modes from which we can get an impression of the size of the perturbation effects.

Methanol may serve as an example. Lees[78] has reported a change in the barrier height for the internal rotation from 376 to 557 cm^{-1} when going from the ground vibrational state to the first excited state of the CH_3 in-plane rocking mode. This result is based on the assumption that the $\alpha^{(gg')}_k$-terms can be neglected, and although this may be an over-simplification, the example clearly shows that the total effect of the α-terms can be extremely large.

Another striking example is carbon suboxide, C_3O_2, which is a typical quasi-linear molecule[79] with a large amplitude CCC-bending vibration, ν_7. The spectroscopic data has been analyzed by Weber and Ford[22] using a simple approximation of the effective bending-rotation operator. In this the four elements of the diagonal $B_{gg}^{(v)}$-matrix are all related to the single parameter B^0, depending on the vibrational state, and the effective bending potential is expressed by two terms, a harmonic and a quartic. It turns out that the bending potential changes drastically with excitation of the antisymmetric C=C stretch mode at 1,587 cm^{-1}, ν_4. Thus the potential hump at the linear configuration is 30.56 cm^{-1} in the g.s., whereas 56.58 cm^{-1} is found in the ν_4 excited state. On the other hand, B^0, which is the rotational constant of the linear configuration, only changes from 0.0735138 to 0.0733140 cm^{-1}. Much smaller effects on the potential are observed by exciting to the $\nu_2 + \nu_3$ state.

Hitherto all predictions on a valence theoretical basis have indicated that C_3O_2 should have a linear equilibrium configuration. It has therefore been a puzzling question why potential humps are actually observed. Considering Eq. (4.58) an obvious suggestion is that the discrepancy is due to the $\alpha_k^{(V)}$-contributions, since we observe $V^{(v)}$ whereas the predicted potential is V^0. From the observations on C_3O_2 it is possible to estimate $\alpha_4^{(V)}$ and $\alpha_2^{(V)} + \alpha_3^{(V)}$, thus allowing a partial correction of the potential function. This gives a reduction in the hump from 30.56 to 22.04 cm^{-1}. It is an exciting question whether α's for the remaining vibrational modes may account for the remainder so that a linear equilibrium configuration will eventually result from experiment as well. By similar arguments it has been possible to show that this is the case in fulminic acid (HCNO)[21].

5 A Case Study, C_3

Many molecular studies of large amplitude internal motions might serve as a basis for an example of how the methods presented above can be applied in practice. C_3 was chosen primarily because it is the simplest one available. This should make it easier to follow the evaluation of the limited number of rather simple terms entering the Hamiltonian.

Although C_3 is a radical, not met under ordinary laboratory conditions, it has been studied with considerable interest. Recently graphite has been considered as an ablasive material for protecting heat exposed surfaces of space vehicles, and since C_3 is one of the main species formed by evaporating graphite, it is essential to know the thermodynamic properties of C_3[80]. In an attempt to estimate the partition function Hansen, Henderson and Pearson[80, 81] have discussed various models for the bending vibration, ν_2, which in the investigation by Gausset, Herzberg, Lagerqvist and Rosen[82] was found to be extremely floppy and anharmonic. Thus they estimated the wave number 63 cm^{-1} for the lowest bending mode, and it was found that the vibrational levels diverge corresponding to a rather broad, but steep potential well. However, no rigorous attempt to treat the combined rotation – large amplitude internal motion has appeared.

The C_3 spectra to be discussed here were obtained by Gausset et al.[82] and Merer[83] in absorption and fluorescense by flash photolysis of diazomethane and diazopropyne mixed with a large extent of inert gas. The transitions form a group around 4,050 Å which has also been observed in cometary spectra. It was established that the transitions are rovibrational components of a $^1\Pi_u - {}^1\Sigma_g^+$ electronic transition, and by forming appropriate combination differences, information about bending states up to $\nu_2 = 6$ in the two electronic states can be obtained. In the $^1\Pi_u$ electronic state a much higher bending frequency, 308 cm^{-1}, was found and therefore a discussion of the much more impeded internal motions of this latter state is irrelevant in the present context.

Only limited information is available about the stretching modes. Weltner and McLeod[84] have investigated the spectra of C_3 trapped in an inert gas matrix. They assigned absorptions at 1,235 cm^{-1} and 2,040 cm^{-1} to the symmetric and antisymmetric stretching modes, ν_1 and ν_3, respectively. The former assignment was confirmed by Merer[83]. He assigned band of the 4,050 Å group to transitions involving the $|1, 0, 0\rangle$ and $|2, 0, 0\rangle$ vibrational states of the electronic ground state from which $\nu_1 = 1224.5$ cm^{-1} and $2\,\nu_1 = 2436$ cm^{-1} were determined. However, on this basis it is impossible to gain any deeper insight into the perturbation effects from the small amplitude motions.

5.1 Reference Structure, I^0 and μ^0 Matrices

As outlined in Sect. 4.9 we must begin an explicit development of the Hamiltonian by considering the semirigid rotor (Sect. 4.2) which serves as the reference.

The reference configurations are given by the constant C,C bond lengths, R, and a bending angle, ρ, defined so that $\rho = 0$ in the linear case (Fig. 1). The molecular axes are chosen so that the y-axis is perpendicular to the molecular plane and with the x-axis bisecting the C,C,C angle of the reference. This choice is obvious considering the symmetry of the reference, C_{2v}.

In Table 1 are given the components of the position vectors, $\mathbf{a}_\alpha$; $\alpha = 1,2,3$, and the four sets of $\mathbf{t}^0_{g,\alpha}$; $g = x, y, z, \rho$, [Eq. (3.8) and (4.3)], which are required for de-

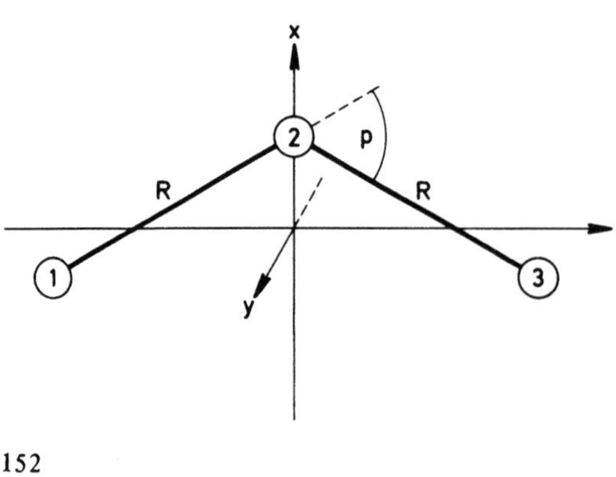

Fig. 1. Reference configuration and molecular axes of C_3

Table 1. Components of vectors, $\mathbf{a}_\alpha$, $\mathbf{t}^0_{g,\alpha}$, $\mathbf{l}_{k,\alpha}$ relevant to the Hamiltonian of C_3

	C_1			C_2			C_3		
	z	x	y	z	x	y	z	x	y
$a_{\alpha g}$	$-R\sqrt{f_2/2}$	$-\frac{R}{3}\sqrt{f_0/2}$	0	0	$\frac{R}{3}\sqrt{2f_0}$	0	$R\sqrt{f_2/2}$	$-\frac{R}{3}\sqrt{f_0/2}$	0
$t^0_{z,\alpha g}$	0	0	$-\frac{R}{3}\sqrt{f_0/2}$	0	0	$\frac{R}{3}\sqrt{2f_0}$	0	0	$-\frac{R}{3}\sqrt{f_0/2}$
$t^0_{x,\alpha g}$	0	0	$R\sqrt{f_2/2}$	0	0	0	0	0	$-R\sqrt{f_2/2}$
$t^0_{y,\alpha g}$	$\frac{R}{3}\sqrt{f_0/2}$	$-R\sqrt{f_2/2}$	0	$-\frac{R}{3}\sqrt{2f_0}$	0	0	$\frac{R}{3}\sqrt{f_0/2}$	$R\sqrt{f_2/2}$	0
$t^0_{\rho,\alpha g}$	$\frac{R}{2}\sqrt{f_0/2}$	$-\frac{R}{6}\sqrt{f_2/2}$	0	0	$\frac{R}{6}\sqrt{2f_2}$	0	$-\frac{R}{6}\sqrt{f_0/2}$	$-\frac{R}{6}\sqrt{f_2/2}$	0
$l_{1,\alpha g}$	$-\frac{1}{2}\sqrt{f_2/f_1}$	$-\frac{1}{2}\sqrt{f_0/f_1}$	0	0	$\sqrt{f_0/f_1}$	0	$\frac{1}{2}\sqrt{f_2/f_1}$	$-\frac{1}{2}\sqrt{f_0/f_1}$	0
$l_{2,\alpha g}$	$-\frac{1}{2}\sqrt{f_2/f_3}$	$-\frac{1}{2}\sqrt{f_0/f_3}$	0	$\sqrt{f_2/f_3}$	0	0	$-\frac{1}{2}\sqrt{f_2/f_3}$	$\frac{1}{2}\sqrt{f_0/f_3}$	0
$\frac{\partial}{\partial\rho}l_{1,\alpha g}$	$\frac{3}{4}\sqrt{f_0/f_1^3}$	$-\frac{1}{4}\sqrt{f_2/f_1^3}$	0	0	$\frac{1}{2}\sqrt{f_2/f_1^3}$	0	$-\frac{3}{4}\sqrt{f_0/f_1^3}$	$-\frac{1}{4}\sqrt{f_2/f_1^3}$	0
$\frac{\partial}{\partial\rho}l_{2,\alpha g}$	$\frac{1}{4}\sqrt{f_0/f_3^3}$	$-\frac{3}{4}\sqrt{f_2/f_3^3}$	0	$-\frac{1}{2}\sqrt{f_0/f_3^3}$	0	0	$\frac{1}{4}\sqrt{f_0/f_3^3}$	$\frac{3}{4}\sqrt{f_2/f_3^3}$	0

riving the $\mathbf{I}^0$ and $\boldsymbol{\mu}^0$ matrix elements. Here the auxiliary functions, f_n; $n = 0,1,2,3$, have been introduced,

$$\begin{aligned} f_0 &= 1 - \cos\rho, & f_1 &= 2 - \cos\rho \\ f_2 &= 1 + \cos\rho, & f_3 &= 2 + \cos\rho \end{aligned} \tag{5.1}$$

Notice, that $f_n = n$ for $\rho = 0$.

From Eq. (4.5) we find that $\mathbf{I}^0$ is always diagonal and therefore the elements of μ^0 are easily obtained,

$$\left.\begin{aligned} \mu^0_{xx} &= 2/If_2, & \mu^0_{yy} &= 3/If_3, \\ \mu^0_{zz} &= 6/If_0, & \mu^0_{\rho\rho} &= 12/If_1, \end{aligned}\right\} \quad I = 2\,mR^2 \tag{5.2}$$

where I is the moment of inertia of the reference in a linear configuration, $\rho = 0$.

5.2 Normal Coordinates and l-Vectors

Defining $\Delta R_1 = \delta(C_1, C_2)$ and $\Delta R_2 = \delta(C_2, C_3)$ as the small amplitude bond stretching coordinates it immediately follows from the symmetry that the normal coordinates can be defined by

$$\begin{aligned} Q_1 &= (\Delta R_1 + \Delta R_2)\sqrt{m/2f_1} \\ Q_2 &= (\Delta R_1 - \Delta R_2)\sqrt{m/2f_3} \end{aligned} \tag{5.3}$$

The corresponding l-vectors of Table 1 were evaluated employing Eqs. (2.31) and (3.45). The normalization factors of Eq. (5.3) may be verified by inspecting the conditions of Eq. (3.46).

5.3 μ-Derivatives and Coriolis Coupling Constants

From the vectors of Table 1 we can now evaluate the quantities appearing in the kinetic part of H up to H_4, [Eq. (4.36)]. The derivatives of μ are found using Eqs. (3.51), (4.20), (4.21) and (4.23), while ζ-functions follows from Eqs. (3.47) and (4.15). The resulting μ-derivatives are given in Table 2, whereas the only nonvanishing ζ is

$$\zeta^y_{12} = -\zeta^y_{21} = -(f_0 f_2/f_1 f_3)^{1/2} \tag{5.4}$$

It must be emphasized that the derivatives of Table 2 are taken with respect to the normal coordinates, Q_1 and Q_2, whereas derivatives with respect to dimensionless coordinates, q_1 and q_2, are required in Eq. (4.36) and subsequent equations. These derivatives are obtained by multiplying with

$$\frac{\partial Q_k}{\partial q_k} = (\hbar/2\,\pi c\omega_k)^{1/2} \tag{5.5}$$

Table 2. Derivatives of **J** and **μ** elements ($a_k^{(gg')} = 2\,J_{gg'}^{(k)}$) with respect to normal coordinates, Q_1 and Q_2

$J_{xx}^{(1)} = \frac{1}{2} f_2 (I/f_1)^{1/2}$	$J_{yy}^{(1)} = (I/f_1)^{1/2}$
$J_{zz}^{(1)} = \frac{1}{2} f_0 (I/f_1)^{1/2}$	$J_{\rho\rho}^{(1)} = \frac{1}{4} (I/f_1)^{1/2}$
$J_{zx}^{(2)} = -\frac{1}{2} (If_0 f_3)^{1/2}$	$J_{y\rho}^{(2)} = \frac{1}{2} (I/f_3)^{1/2}$
$\mu_{xx}^{(1)} = -4 (I^3 f_1)^{-1/2} f_2^{-1}$	$\mu_{yy}^{(1)} = -18 (I^3 f_1)^{-1/2} f_3^{-2}$
$\mu_{zz}^{(1)} = -36 (I^3 f_1)^{-1/2} f_0^{-1}$	$\mu_{\rho\rho}^{(1)} = -72 (If_1)^{-3/2} f_1^{-1}$
$\mu_{zx}^{(2)} = 12 (I^3 f_0 f_2 f_3)^{-1/2}$	$\mu_{y\rho}^{(2)} = -36 (If_3)^{-3/2} f_1^{-1}$
$\mu_{xx}^{(11)} = 12 (I^2 f_1 f_2)^{-1}$	$\mu_{xx}^{(22)} = 36 (I^2 f_2 f_3)^{-1}$
$\mu_{yy}^{(11)} = 162 (I^2 f_1 f_3^3)^{-1}$	$\mu_{yy}^{(22)} = 162 (I^2 f_1 f_3^3)^{-1}$
$\mu_{zz}^{(11)} = 324 (I^2 f_0 f_1)^{-1}$	$\mu_{zz}^{(22)} = 108 (I^2 f_0 f_3)^{-1}$
$\mu_{\rho\rho}^{(11)} = 648 (I^2 f_1^4)^{-1}$	$\mu_{\rho\rho}^{(22)} = 648 (I^2 f_1^2 f_3^2)^{-1}$

once or twice for first and second derivatives respectively. Derivatives of $B_{gg'}$ require the additional factor $\hbar/4\pi c$, e.g.

$$B_{xx}^{(1)} = \frac{\partial B_{xx}}{\partial q_1} = \frac{h}{8\pi^2 c} \left(\frac{h}{4\pi^2 c\omega_1}\right)^{1/2} \frac{\partial \mu_{xx}}{\partial Q_1} = -\frac{4B}{f_2} \left(\frac{2B}{\omega_1 f_1}\right)^{1/2}$$
$$B = h/8\pi^2 cI \tag{5.6}$$

The derivatives thus obtained are given in Table 3.

Table 3. Derivatives of $B_{gg'}$ elements with respect to dimensionless normal coordinates, $q_k = (2\pi c\omega_k/\hbar)^{1/2} Q_k$

$B_{xx}^{(1)} = -4B (2B/\omega_1 f_1)^{1/2} f_2^{-1}$	$B_{yy}^{(1)} = -18B (2B/\omega_1 f_1)^{1/2} f_3^{-2}$
$B_{zz}^{(1)} = -36B (2B/\omega_1 f_1)^{1/2} f_0^{-1}$	$B_{\rho\rho}^{(1)} = -72B (2B/\omega_1 f_1)^{1/2} f_1^{-2}$
$B_{zx}^{(2)} = 12B/(2B/\omega_2 f_0 f_2 f_3)^{1/2}$	$B_{y\rho}^{(2)} = -36B (2B/\omega_2 f_3)^{1/2} f_1^{-1} f_3^{-1}$
$B_{xx}^{(11)} = 24B^2/\omega_1 f_1 f_2$	$B_{xx}^{(22)} = 72B^2/\omega_2 f_2 f_3$
$B_{yy}^{(11)} = 324B^2/\omega_1 f_1 f_3^3$	$B_{yy}^{(22)} = 324B^2/\omega_2 f_1 f_3^3$
$B_{zz}^{(11)} = 648B^2/\omega_1 f_0 f_1$	$B_{zz}^{(22)} = 216B^2/\omega_2 f_0 f_3$
$B_{\rho\rho}^{(11)} = 1296B^2/\omega_1 f_1^4$	$B_{\rho\rho}^{(22)} = 1296B^2/\omega_2 f_1^2 f_3^2$

5.4 Anharmonic Force Constants and α-Functions

The available experimental data are insufficient for a determination of the parameters, Φ_{111} and Φ_{122}, entering the expression for the α-functions [Eq. (4.59)]. However, Hoy et al. [6] have discussed how the relations can be formed between these parameters and the force constants of an expansion of the potential energy in the true valence coordinates, $\bar{\rho}$, $\mathcal{R}_1$ and $\mathcal{R}_2$,

$$V = V(\bar{\rho}) + \sum_i \frac{\partial V(\bar{\rho})}{\partial \mathcal{R}_i} \mathcal{R}_i + \frac{1}{2} \sum_{ij} f_{ij}(\bar{\rho})\, \mathcal{R}_i \mathcal{R}_j + \frac{1}{6} \sum_{ijk} f_{ijk}(\bar{\rho})\, \mathcal{R}_i \mathcal{R}_j \mathcal{R}_k + \ldots \tag{5.7}$$

Introducing reasonable approximations in this expression it may then be possible to obtain fair estimates of the Φ's.

First we shall neglect the difference between the true bending angle, $\bar{\rho}$, and the angle, ρ, of the reference configuration in all terms except the leading term of $V(\bar{\rho})$,

$$\frac{1}{hc} V(\bar{\rho}) = V_2 \rho^2 + V_4 \rho^4 + V_6 \rho^6 + V_2(\bar{\rho}^2 - \rho^2) \tag{5.8}$$

In this way we may still get an impression of the importance of this difference.

Secondly, we approximate the second term of Eq. (5.7) by

$$\sum_i \frac{\partial V(\rho)}{\partial \mathcal{R}_i} \mathcal{R}_i = \frac{1}{2} f_{r\rho\rho}\, \rho^2 (\mathcal{R}_1 + \mathcal{R}_2) = hc\, V_2^r \rho^2 (\mathcal{R}_1 + \mathcal{R}_2) \tag{5.9}$$

where $f_{r\rho\rho} = 2\, hc\, V_2^r$ is constant. Finally the small amplitude contributions to V are approximated by

$$V^{(2)} + V^{(3)} = \frac{1}{2} f_{rr} (\mathcal{R}_1^2 + \mathcal{R}_2^2) + f_{rr'}\, \mathcal{R}_1 \mathcal{R}_2 + \frac{1}{6} f_{rrr} (\mathcal{R}_1^3 + \mathcal{R}_2^3) \tag{5.10}$$

The harmonic force constants, f_{rr} and $f_{rr'}$, are identical to those found when rectilinear coordinates are used. They may therefore be estimated from the observed stretching frequencies using $\lambda_k = (2\pi c\omega_k)^2$ and Eq. (5.3),

$$\begin{aligned} f_{rr} &= \lambda_1 m/2 f_1 + \lambda_2 m/2 f_3 \approx 10.3 \text{ mdyn/Å} \\ f_{rr'} &= \lambda_1 m/2 f_1 - \lambda_2 m/2 f_3 \approx 0.5 \quad - \end{aligned} \tag{5.11}$$

As pointed out by Kuchitsu and Morino[53], it is a fair approximation to neglect all force constants of $V^{(3)}$ except the stretching constant, f_{rrr}. The order of magnitude of this can be estimated assuming a Morse potential for the individual C–C stretchings,

$$V(\mathcal{R}_i) = D_e(1 - \exp(-a\,\mathcal{R}_i))^2, \; i = 1, 2 \tag{5.12}$$

giving

$$f_{rr} = D_e a^2, \; f_{rrr} = -3\,a f_{rr}, \; f_{rrrr} = 7\,a^2 f_{rr} \tag{5.13}$$

The parameter, a, has been determined for a large number of diatomic and a few three-atom molecules. Remarkably it appears that the values are closely scattered around $a = 2$ Å^{-1}, and we may therefore estimate f_{rrr} and f_{rrrr} using this a-value.

Among the spectroscopic constants only x_{kk} depends on the quartic constants. Since the experimental data are very limited for the states with an excited stretching mode, it is impossible to obtain any closer check on these constants which are therefore left out in the following discussion.

Using Eqs. (22), (23), (25) and (26) from Ref.[6] we can evaluate the elements of the nonlinear transformation,

$$\mathcal{R} = \mathbf{L}^*(\rho)\,\mathbf{Q}, \tag{5.14}$$

here to second order only, since we are neglecting the quartic constants. The expansion coefficients are given in Table 4.

Substituting Eq. (5.14) into Eq. (5.7) and introducing dimensionless normal coordinates, the potential function becomes,

$$\begin{aligned} V/hc = {} & V_2\rho^2 + V_4\rho^4 + V_6\rho^6 + V_2^{(1)}\rho^2 q_1 \\ & + \frac{1}{2}\,\omega_1 q_1^2 + \frac{1}{2}\,\omega_2 q_2^2 + \frac{1}{6}\,\Phi_{111}\,q_1^3 + \frac{1}{2}\,\Phi_{122}\,q_1 q_2^2 \end{aligned} \tag{5.15}$$

where,

$$\begin{aligned} V_2^{(1)} & = (R\,V_2' + V_2 f_1^{-1/2})(8\,B/\omega_1)^{1/2} \\ \Phi_{111} & = \frac{3\,f_0 f_2}{f_1}\left(\frac{2\,B\omega_1}{f_1}\right)^{1/2} - \left(1 + \frac{\omega_2^2 f_1}{\omega_1^2 f_3}\right)(\kappa\,\omega_1 f_1)^{1/2} \\ \Phi_{122} & = \frac{f_0 f_2}{f_3}\,\omega_2\left(\frac{\omega_1^2}{\omega_2^2} - 2\right)\left(\frac{2\,B}{\omega_1 f_1}\right)^{1/2} - \left(1 + \frac{\omega_1^2 f_3}{\omega_2^2 f_1}\right)\omega_2\left(\frac{\kappa f_1}{\omega_1}\right)^{1/2} \\ \kappa & = \frac{9\,ha^2}{32\,\pi^2\,cm} = (a/2)^2\;12.643\ \text{cm}^{-1} \end{aligned} \tag{5.16}$$

Table 4. First and second derivatives of true valence coordinates, $\mathcal{R}_1$, $\mathcal{R}_2$ and $\bar{\rho} - \rho$ with respect to normal coordinates, Q_1 and Q_2

$L_1^1 = L_2^1 = (f_1/2\,m)^{1/2}$	$L_1^2 = -L_2^2 = (f_3/2\,m)^{1/2}$
$L_1^{11} = L_2^{11} = f_0 f_2/2\,mRf_1$	$L_1^{22} = L_2^{22} = f_0 f_2/2\,mRf_3$
$L_1^{12} = -L_2^{12} = -f_0 f_2/2\,mR(f_1 f_3)^{1/2}$	
$L_\rho^1 = (2\,f_0 f_2/mf_1)^{1/2}/R$	$L_\rho^2 = 0$
$L_\rho^{11} = -L_\rho^{22} = -2\,(f_0 f_2)^{1/2}/mR^2$	$L_\rho^{12} = 0$

The order of magnitude of V_2 is 200 cm^{-1}, assuming a harmonic bending motion with $\omega = 63$ cm^{-1} and $B = 0.42$ cm^{-1}. If we neglect the relaxation term V_2^r, we find an approximate value, $V_2^{(1)} = 10$ cm^{-1}, originating from the difference $\overline{\rho} - \rho$. Similarly the Φ's are estimated,

$$\begin{aligned} \rho = 0\text{:} \quad & \Phi_{111} = -239 \text{ cm}^{-1}, \ \Phi_{122} = -433 \text{ cm}^{-1} \\ \dot{\rho} = 60°\text{:} \ & \Phi_{111} = -364 \text{ cm}^{-1}, \ \Phi_{122} = -429 \text{ cm}^{-1} \end{aligned} \tag{5.17}$$

indicating that the contributions to the "rotational constants" from $V_2^{(1)}$ can be neglected compared to the considerably larger Φ's. However, as pointed out (Sect. 4.8.2.3) we must consider the possibility that V_2^r may be underestimated.

Table 5. Functions, $\alpha_k^{(gg)}$

k, gg	$-\alpha_k^{(gg)}/B_{gg}^0$
1, xx	$\dfrac{6B}{\omega_1 f_1}(1 + f_0 f_2/f_1) - \gamma_1$
1, yy	$\dfrac{6B}{\omega_1 f_1 f_3^2}\left(9 + f_0 f_2 \dfrac{3\omega_1^2 + \omega_2^2}{\omega_1^2 - \omega_2^2}\right) + \dfrac{18 B f_0 f_2}{\omega_1 f_1^2 f_3} - \dfrac{3\gamma_1}{f_3}$
1, zz	$\dfrac{18B}{\omega_1 f_1}(f_3 + 3 f_0/f_1) - 3\gamma_1$
1, $\rho\rho$	$\dfrac{18 B f_3}{\omega_1 f_1^2} - \dfrac{3\gamma_1}{f_1}$
2, xx	$\dfrac{18B}{\omega_2 f_3} + \beta - \gamma_2$
2, yy	$\dfrac{6B}{\omega_2 f_1 f_3^2}\left(9 + f_0 f_2 \dfrac{3\omega_2^2 + \omega_1^2}{\omega_2^2 - \omega_1^2}\right) + 3(\beta - \gamma_2)/f_3$
2, zz	$\dfrac{18B}{\omega_2 f_3} + 3(\beta - \gamma_2)$
2, $\rho\rho$	$\dfrac{54B}{\omega_2 f_1 f_3^2} + 3(\beta - \gamma_2)/f_1$

$$\beta = \frac{2 B f_0 f_2}{\omega_2 f_1 f_3}(1 - 2\omega_2^2/\omega_1^2)$$

$$\gamma_1 = \frac{(2B\kappa)^{1/2}}{\omega_1}(1 + \omega_2^2 f_1/\omega_1^2 f_3)$$

$$\gamma_2 = \frac{(2B\kappa)^{1/2}}{\omega_2}(\omega_2^2/\omega_1^2 + f_3/f_1) = \gamma_1 \omega_1 f_3/\omega_2 f_1$$

Since Φ_{111} varies noticeably with ρ, the functional form given in Eq. (5.16) is retained when evaluating the α's, rather than using constant values, $\Phi^0_{kkk'}$, as assumed in the perturbation treatment of Sect. 4.8.

The expression of $\alpha_k^{(gg')}$, [Eq. (4.59)], can be rewritten as

$$-\alpha_k^{(gg')} = \frac{1}{2} B_{gg'}^{(kk)} - \sum_l B_{gg'}^{(l)} \frac{\Phi_{kkl}}{2\,\omega_l} + 2\, B^0_{gy} B^0_{yg'} \sum_l (\zeta^y_{kl})^2 \frac{3\,\omega_k^2 + \omega_l^2}{\omega_k(\omega_k^2 - \omega_l^2)} \tag{5.18}$$

where it has been considered that only ζ^y_{kl} is nonvanishing. Using Eqs. (5.4), (5.16) and Table 3, we then obtain the α-functions given in Table 5.

5.5 Effective Bending-Rotation Operator

In Eq. (4.32) we shall provisionally neglect the centrifugal distortion terms. $D_J = 4\,B^3/\omega_1^2$ is of the order 1.9×10^{-7} cm^{-1} giving a correction of only 0.18 cm^{-1} for a $\Delta J = 2$ transition when $J = 50$. The effective operator may then be expressed by

$$\frac{1}{hc} H^{(v)} = B(\rho)\, J^2 + b(\rho)\,(J_x^2 - J_y^2) + J_\rho\, C(\rho)\, J_\rho + C(0)\left(J_z^2 - \frac{1}{4}\right) \rho^{-2} + A(\rho)\, J_z^2 + V(\rho) \tag{5.19}$$

where superscripts (v) have been omitted from the functions $B(\rho)$, $b(\rho)$, etc. For the g.s. of the stretchings these functions are given by

$$\begin{aligned} B(\rho) &= \frac{1}{2}(B^0_{xx} + B^0_{yy}) - \frac{1}{4}(\alpha_1^{(xx)} + \alpha_2^{(xx)} + \alpha_1^{(yy)} + \alpha_2^{(yy)}) \\ b(\rho) &= \frac{1}{2}(B^0_{xx} - B^0_{yy}) - \frac{1}{4}(\alpha_1^{(xx)} + \alpha_2^{(xx)} - \alpha_1^{(yy)} - \alpha_2^{(yy)}) \\ C(\rho) &= B^0_{\rho\rho} - \frac{1}{2}(\alpha_1^{(\rho\rho)} + \alpha_2^{(\rho\rho)}) \\ A(\rho) &= B^0_{zz} - \frac{1}{2}(\alpha_1^{(zz)} + \alpha_2^{(zz)}) - C(0)/\rho^2 - B(\rho) \end{aligned} \tag{5.20}$$

It should be noted that the only singularity appears at $\rho = 0$ in the fourth term of $H^{(v)}$ [Eq. (5.19)] due to the factor ρ^{-2}. This causes no trouble, however, since the term enters the two-dimensional harmonic oscillator operator, H^0_b, used below [Eq. (5.22)] in defining a basis for the matrix representation of $H^{(v)}$. Also note that the term contains a contribution $-C(0)/4\,\rho^2$ which has its origin in the U-function [Eqs. (2.85)–(2.87)]. From these equations it can be shown that $U_1 + U_2$ is given by the same simple expression as the U-function for a rigid nonlinear molecule [Eq. (3.84)], whereas U_3 appears to be small. The singularity in U can be appropriately taken care of by moving the term $-C(0)/4\,\rho^2$ from U to H_b^0. The remaining part of U can be considered as a small perturbation of the effective potential $V(\rho)$. It is emphasized that $A(\rho)$ remains finite at $\rho = 0$,

$$A(0)/B = -30\,B/\omega_1 + 6\,B/\omega_2(3 - 4\,\omega_2^2/\omega_1^2) + (2\,B\kappa)^{1/2}(33/\omega_2 - \omega_2/\omega_1^2 - 1/\omega_1 - 13\,\omega_2^2/3\,\omega_1^3) \quad (5.21)$$

5.6 Calculation of Energies

The commutation relations for the momenta and the coordinate, ρ, appearing in $H^{(v)}$ are

$$[J_f, J_g] = -i\,J_h;\ (f, g, h) = (x, y, z)_{\text{cyclic}}$$
$$[J_f, J_\rho] = 0,\ [J_f, \rho] = 0,\ [J_\rho, \rho] = -i \quad (5.22)$$

Notice, that the relations for the angular momenta are the same as those applying to nonlinear molecules. This is a consequence of our definition of a molecular coordinate system with three degrees of freedom.

From Eq. (5.22) it follows that $H^{(v)}$ commutes with J^2 and J_Z and consequently $H^{(v)}$ is diagonal in the J and M quantum numbers. Hence the eigenvalues are independent of M and can be found for individual J-values, either by diagonalizing the Hamilton matrix, formed in an appropriate basis, or by numerical integration. The former method has been applied below.

5.6.1 Basis Vectors and Matrix Elements

Basis vectors are defined as the simultaneous eigenvectors of J^2, J_Z, J_z and H_b^0, where

$$H_b^0 = C(0)\left(J_\rho^2 + \left(J_z^2 - \frac{1}{4}\right)/\rho^2\right) + D\rho^2 \quad (5.23)$$

is the operator of a two-dimensional harmonic oscillator[60]. The constant D may be adjusted so that the spacing between the eigenvalues of H_b^0 is optimized in relation to the eigenvalues of $H^{(v)}$.

A basis vector is designated by $|n, l, J, M\rangle$ in accordance with

$$J^2|n, l, J, M\rangle = |n, l, J, M\rangle J(J+1)$$
$$J_Z|n, l, J, M\rangle = |n, l, J. M\rangle M$$
$$J_z|n, l, J, M\rangle = |n, l, J, M\rangle l \quad (5.24)$$
$$H_b^0|n, l, J, M\rangle = |n, l, J, M\rangle E^0(n+1)$$

The two first equations follow from the commutation relations of the angular momentum components, J_F; $F = X, Y, Z$, along space fixed axes. The latter two are not as obvious as they may appear at first glance. Since J_x and J_y do not commute with H_b^0 we cannot factorize the vector space to treat H_b^0 and J_z in a separated basis of vectors, $\{|n, l\rangle\}$, such as it is usually assumed when discussing the two-dimensional harmonic oscillator. $J_y + iJ_x$ cannot be used as ladder operators of J_z, and similarly it may be shown that the usual ladder operators[85)] for H_b^0 are inapplicable as well, since they do not commute with J^2.

However, from the commutation relations [Eq. (5.22)], it follows that the two operators, L_+ and L_-,

$$
\begin{aligned}
L_\pm &= \left(\alpha\rho - \frac{i}{\alpha} J_\rho \pm \frac{1}{\alpha\rho} J_z - \frac{1}{2\,\alpha\rho}\right)(J_y \pm iJ_x) \\
&= (J_y \pm iJ_x)\left(\alpha\rho - \frac{i}{\alpha} J_\rho \pm \frac{1}{\alpha\rho} J_z + \frac{1}{2\,\alpha\rho}\right) \\
\alpha^4 &= D/C,
\end{aligned}
\tag{5.25}
$$

and their adjoint operators are the relevant ladder operators, commuting with J_F and J^2, and with

$$[H_b^0, L_\pm] = E^0 L_\pm,\ [J_z, L_\pm] = \pm L_\pm \tag{5.26}$$

where

$$E^0 = 2\,D/\alpha^2 = 2\,C\alpha^2 = 2\sqrt{CD} \tag{5.27}$$

From Eq. (5.26) it is seen that the eigenvalues of H_b^0 and J_z are equally spaced, with spacings E^0 and 1 respectively, in accordance with Eq. (5.24). Furthermore

$$
\begin{aligned}
L_+L_+^\dagger &= 2\,(J^2 - J_z^2 + J_z)(H_b^0/E^0 + J_z - 1) \\
L_-L_-^\dagger &= 2\,(J^2 - J_z^2 - J_z)(H_b^0/E^0 - J_z - 1) \\
L_+^\dagger L_+ &= 2\,(J^2 - J_z^2 - J_z)(H_b^0/E^0 + J_z + 1) \\
L_-^\dagger L_- &= 2\,(J^2 - J_z^2 + J_z)(H_b^0/E^0 - J_z + 1)
\end{aligned}
\tag{5.28}
$$

from which it may be deduced that the quantum numbers, n and l, must be integers with the usual restrictions,

$$
\begin{aligned}
&n \in \{0, 1, 2, \ldots\} \\
&l \in \{n, n-2, \ldots -n\} \text{ and } l < J
\end{aligned}
\tag{5.29}
$$

Finally one finds that the matrix elements of L_+ and L_-, omitting the quantum number M, can be given by

$$
\begin{aligned}
\langle n, l, J|L_+|n-1, l-1, J\rangle &= p_+[2\,(n+l)\,(J(J+1) - l\,(l-1))]^{1/2} \\
\langle n, l, J|L_-|n-1, l+1, J\rangle &= p_-[2\,(n-l)\,(J(J+1) - l\,(l+1))]^{1/2}
\end{aligned}
\tag{5.30}
$$

where p_+ and p_- are phase factors of modulus one subject to the restriction $p_+p_-^* = \pm 1$ [Eq. (5.40)].

From Eqs. (5.24) and (5.30) all relevant matrix elements can be evaluated. Thus as an example we may consider

$$L_+ + L_-^\dagger = 2\,\alpha\rho\,(J_y + \mathrm{i}J_x) \tag{5.31}$$

from which we obtain

$$\frac{1}{4}(L_+ + L_-^\dagger)(L_+^\dagger + L_-) = \alpha^2\rho^2(J^2 - J_z^2 + J_z) \tag{5.32}$$

$$\frac{1}{8}(L_+ + L_-^\dagger)^2 + \frac{1}{8}(L_+^\dagger + L_-)^2 = -\alpha^2\rho^2(J_x^2 - J_y^2) \tag{5.33}$$

Applying Eq. (5.30) to the left side of Eq. (5.32) it follows that the matrix elements of ρ^2 can be found isolated, since the same J, l-factors turn up on both sides of the equation,

$$\begin{aligned} \langle n, l, J \mid \alpha^2\rho^2 \mid n, l, J\rangle &= n + 1 \\ \langle n, l, J \mid \alpha^2\rho^2 \mid n-2, l, J\rangle &= \frac{1}{2}p_+p_-[(n+l)(n-l)]^{1/2} \end{aligned} \tag{5.34}$$

This is the same result as found for an isolated two-dimensional oscillator[64]. But the matrix elements found in a similar way from Eq. (5.33) are not as simply related to those obtained in the usual case,

$$\langle n, l, J \mid \alpha^2\rho^2(J_x^2 - J_y^2) \mid n, l-2, J\rangle = -p_+p_-^*\,[(n+l)(n-l+2)f(J, l-1)]^{1/2}$$

$$\langle n, l, J \mid \alpha^2\rho^2(J_x^2 - J_y^2) \mid n-2, l\pm 2, J\rangle = -\frac{1}{2}p_\mp^2[(n\mp l)(n\mp l-2)f(J, l\pm 1)]^{1/2}$$

$$f(J,k) = \frac{1}{4}[J(J+1) - k(k+1)][J(J+1) - k(k-1)] \tag{5.35}$$

A formal relation between these and corresponding formulae for an isolated oscillator may be expressed by including the matrix elements of a rigid rotor,

$$\begin{aligned} &\langle n, l, J \mid \rho^2(J_x^2 - J_y^2) \mid n', l', J\rangle = \\ &\langle n, l \mid 2\rho^2\cos(2\chi) \mid n', l'\rangle\langle J, l \mid J_x^2 - J_y^2 \mid J, l'\rangle \end{aligned} \tag{5.36}$$

where χ is the polar angle of the oscillator[64], and l replaces the usual K-quantum number in the second factor.

5.6.2 Hamilton Matrix

To form the matrix elements of $H^{(\nu)}$ [Eq. (5.19)], it is necessary to expand the various functions of ρ in power series. This has been done numerically by fitting even polynomials to the values of the functions calculated for $\rho = 0°, 10°, 20°, \dots 80°$. Bending angles as large as 80° should be considered since the classical vibration amplitudes of the investigated states are estimated up to about 70°. This made it necessary to include up to ρ^6-terms in the expansions (Table 6).

The only operator nondiagonal in l is $b(\rho)(J_x^2 - J_y^2)$. It contributes essentially only in case of l-type doubling and l-type resonance. The energies are therefore most easily calculated by considering first a Hamiltonian, H', from which this term has been removed. The corresponding matrix is then diagonal in l as well as J, and the individual J, l-blocks can be separately diagonalized. The eigenvectors and eigenvalues of H' are denoted according to

$$H' | \nu, l, J \rangle = | \nu, l, J \rangle E'_{\nu,l,J} \qquad \nu = |l|, |l|+2, \dots ; \; J \geqslant |l| \tag{5.37}$$

The energy levels are doubly degenerate for $|l| > 0$, $E'_{\nu,l,J} = E'_{\nu,-l,J}$ but these degeneracies are lifted by the couplings produced by the term $b(\rho)(J_x^2 - J_y^2)$. The couplings between levels with different quantum number ν can be neglected, however, so that the final energies can be found merely by a second diagonalization of a small matrix of dimension $\nu + 1$ with diagonal elements $E'_{\nu,l,J}$; $l = \nu, \nu - 2, \dots - \nu$, and off-diagonal elements

$$\langle \nu, l, J | b(\rho)(J_x^2 - J_y^2) | \nu, l', J \rangle$$

From symmetry considerations it follows that this matrix, by a similarity transformation, can be separated into two diagonal blocks of dimension $\nu/2$ and $\nu/2 + 1$ for even ν and $(\nu + 1)/2$, both, for odd ν.

Table 6. Expansion coefficients, a_i, for the ρ-dependent functions of $H^{(0)}$, with mean deviation, σ, of fit. $B = 0.4191$ cm^{-1} and $a = 2$ Å^{-1} were assumed in calculating $\alpha_k^{(gg)}$ contributions. A relaxation parameter, k (see Text), was assumed equal to 0 or -0.002 in evaluating set 1 and 2 respectively

		a_0	a_2	a_4	a_6	σ
$B(\rho)/B$	(1)	0.994509	0.208556	0.024076	0.005200	$6 \cdot 10^{-5}$
	(2)	0.994508	0.206564	0.023956	0.005082	
$b(\rho)/B$	(1)	0	0.042842	0.011200	0.004941	$8 \cdot 10^{-5}$
	(2)	0	0.042840	0.011287	0.004936	
$C(\rho)/12B$	(1)	0.987344	−0.464042	0.187602	−0.034293	$2 \cdot 10^{-3}$
	(2)	0.987301	−0.468941	0.190939	−0.035063	
$A(\rho)/B$	(1)	−0.005582	−0.199494	−0.002592	−0.009319	$1 \cdot 10^{-4}$
	(2)	−0.077562	−0.186106	−0.007500	−0.008388	

A detailed discussion of the symmetry properties of the state vectors is outside the frame of the present example. From the form of $H^{(v)}$ it is evident that the symmetry group is the same as that applying to a rigid rotor[49–52, 66 or 67], i.e. the Four Group,

$$V(x, y, z) = \{E, C_2^x, C_2^y, C_2^z\} \tag{5.38}$$

As for a rigid rotor, symmetry adapted eigenvectors of H' are formed by the Wang transformation,

$$\begin{aligned} |v, l, J, \gamma\rangle &= \{|v, l, J\rangle + (-1)^\gamma |v, -l, J\rangle\}/\sqrt{2} \\ & \qquad l > 0, \gamma = 0 \text{ or } 1 \\ |v, 0, J, 0\rangle &= |v, 0, J\rangle \end{aligned} \tag{5.39}$$

and these vectors may be assigned to the irreducible representations according to the parities of l, J and γ. This means that the matrix elements of $b(\rho)(J_y^2 - J_y^2)$ are diagonal blocks as described above.

Another feature of the transformed coupling matrix is that diagonal elements appear when $l = 1$, since Eq. (5.35) gives

$$\langle v, 1, J, \gamma | b(\rho)(J_x^2 - J_y^2) | v, 1, J, \gamma\rangle = -\frac{1}{2}(-1)^\gamma J(J+1)\langle v, 1 | b(\rho) | v, 1\rangle \tag{5.40}$$

provided $p_+ p_-^* = 1$ is chosen for the phase factors of Eq. (5.30). It is these diagonal elements that give rise to l-type doubling corresponding to the two possible parities of γ, whereas the simplest case of l-type resonance occurs between the two levels with $v, \gamma = 2,0$, $l = 0$ and 2.

5.6.3 Scaling and Truncation of the Basis

In defining the harmonic oscillator basis by Eqs. (5.23)–(5.25) the problem of scaling was postponed. Equation (5.27) shows that the three yet undefined parameters, D, E^0 and α, are interrelated so that they are all determined when anyone has been given a value. A reasonable estimate is most easily obtained for the spacing E^0, which should be close to the mean spacing of the levels considered, in order to minimize the dimensions of the Hamilton matrix. Thus, in the present example it was found that with $E^0 = 80\ \text{cm}^{-1}$ the basis could safely be truncated at $n = 39$ corresponding to matrix dimensions 20 x 20 for the diagonal blocks of H'.

5.7 Numerical Results

In Ref.[82] and [83] the experimental results were analyzed in terms of pure bending frequencies and effective rotational constants. Similarly we may calculate the eigenvalues of $H^{(v)}$ for $J = 0$, i.e. by omitting the two leading terms in Eq. (5.20), and calculate expectation values of $B(\rho)$, giving the effective rotational constants B_{eff},

Table 7. Potential and rotational constants used in test calculations

	I	II	III
V_2, cm^{-1}	144.0536	145.7329	138.3849
V_4, cm^{-1}	80.9077	75.5374	95.8967
V_6, cm^{-1}	66.1082	71.9498	57.9710
B, cm^{-1}	0.41901	0.41903	0.41924
a, $Å^{-1}$	2	2	2
D_J, cm^{-1}	0	$1.93 \cdot 10^{-7}$	$1.93 \cdot 10^{-7}$
k	0	0	−0.002

Table 8. Rotational and vibrational constants (cm^{-1}) in the electronic and stretching mode g.s. of C_3

Bending level v	$G(v)$ obs.	$G(v)$ calc.	$B_{eff.}$ obs.	$B_{eff.}$ calc.	q_v obs.	q_v calc.
0 Σ_g^+	–	0	0.4305	0.4305		
1 Π_u	–	63.05	0.4421	0.4426	0.0055	0.0057
2 Σ_g^+	132.7_2	132.68	0.4519	0.4520		
Δ_g	131.4_4	131.29		0.4535		
3 Π_u	207.2_7	206.97	0.4600	0.4611	0.0112	0.0104
4 Σ_g^+	286.5_2	286.70	0.4675	0.4688		
6 Σ_g^+	458.2	458.32	0.4807	0.4833		

and of $b(\rho)$ giving l-type doubling constants, q_v, for $l = 1$. The values calculated from the parameters given as set I in Table 7 are compared with the experimentally derived quantities in Table 8. The agreement is excellent, considering the approximations involved when analyzing in terms of an effective rotation operator.

However, the theory may be subject to a more serious examination if we calculate the J-dependence of the energy levels directly from the Hamilton matrix as outlined in Sect. 5.6.2, i.e. we may diagonalize the individual J, l-blocks of H' and subsequently treat the l-type resonances by diagonalizing smaller matrices diagonal in v.

Only few data are available for the bending states with v greater than three. These have not been considered here, since we hereby gain the advantage that a second diagonalization only involves a 2 x 2 matrix, with only a single coupling element, e.g. for $v = 2$, $\gamma = 0$

$$\langle 2, J, 0, 0 | b(\rho)(J_x^2 - J_y^2) | 2, J, 2, 0 \rangle$$

to be evaluated.

The results of three such treatments are compared with experimental values in Table 9, p. 166. The observed energy differences given here were determined by forming appropriate combination differences between the directly observed electronic

Table 9. Energy differencies (cm^{-1}) in the electronic g.s. ($^1\Sigma_g^+$) of C_3

J	$^QS_{0,0,0}(J)$	obs. − calc.			$^QS_{0,2,0}(J)$	obs. − calc.			$^QS_{2,2,0}(J)$	obs. − calc.		
	obs.	I	II	III	obs.	I	II	III	obs.	I	II	III
2	6.05	0.02	0.02	0.02								
4	9.49	0.02	0.02	0.02	9.84	−0.13	−0.12	−0.11				
6	12.92	0.01	0.01	0.01	13.74	0.12	0.13	0.14				
8	16.35	0.00	0.00	0.00	17.33	0.04	0.06	0.08				
10	19.81	0.02	0.03	0.02	20.93	−0.03	−0.01	−0.01	20.30	−0.34	−0.32	−0.33
12	23.22	0.00	0.01	0.00	24.60	−0.03	0.00	0.02	23.85	−0.34	−0.31	−0.32
14	26.65	0.00	0.01	0.00	28.14	−0.14	−0.11	−0.08	27.51	−0.22	−0.19	−0.20
16	30.10	0.02	0.03	0.03	31.87	−0.05	−0.01	0.01	31.17	−0.11	−0.07	−0.07
18	33.55	0.04	0.06	0.05	35.64	0.09	0.14	0.16	34.81	−0.01	0.03	0.03
20	36.92	−0.01	0.02	0.01	39.18	0.02	0.07	0.09	38.30	−0.07	−0.01	−0.01
22	40.32	−0.03	0.00	−0.01	42.77	0.01	0.07	0.08	41.76	−0.15	−0.09	−0.08
24	43.76	0.00	0.04	0.02	46.32	−0.04	0.04	0.05	45.38	−0.06	0.01	0.02
26	47.13	−0.04	0.00	−0.01	49.87	−0.07	0.02	0.03	48.82	−0.15	−0.07	−0.06
28	50.55	−0.03	0.03	0.01	53.34	−0.17	−0.07	−0.06	52.26	−0.23	−0.14	−0.13
30	53.93	−0.05	0.02	0.00	56.97	−0.09	0.02	0.03	55.90	−0.11	0.00	0.01
32	57.32	−0.06	0.02	0.00	60.49	−0.12	0.01	0.02	59.49	−0.03	0.09	0.10
34	60.72	−0.05	0.04	0.02					62.89	−0.13	0.00	0.01
36	64.06	−0.10	0.00	−0.02	67.61	−0.06	0.10	0.11	66.37	−0.15	0.00	0.01
38	67.47	−0.08	0.04	0.01	71.01	−0.17	0.01	0.01	69.82	−0.19	−0.02	−0.01
40	70.80	−0.13	0.00	−0.03					73.22	−0.28	−0.08	−0.08
42	74.13	−0.19	−0.03	−0.06								
44	77.52	−0.17	0.00	−0.03					80.12	−0.32	−0.09	−0.08
46	80.89	−0.18	0.01	−0.02								
48	84.22	−0.22	−0.01	−0.04								

Table 9 (continued)

J	$^{Q}Q_{0,0,2}(J)$ obs.	obs. − calc. I	II	III	$^{S}Q_{0,0,2}(J)$ obs.	obs. − calc. I	II	III
4	133.29	0.17	0.18	0.11				
6	133.60	−0.02	0.00	−0.07				
8	134.33	0.00	0.02	−0.02				
10	135.27	0.00	0.04	0.02	133.83	0.27	0.28	0.44
12	136.38	−0.06	−0.01	0.00	134.41	−0.01	0.02	0.18
14	137.74	−0.11	−0.03	0.00	135.16	−0.22	−0.18	−0.02
16	139.31	−0.16	−0.07	−0.01	136.15	−0.31	−0.24	−0.09
18	141.11	−0.20	−0.08	0.01	137.42	−0.24	−0.15	0.01
20	143.13	−0.22	−0.07	0.04	138.74	−0.24	−0.12	0.05
22	145.29	−0.29	−0.11	0.03	140.15	−0.27	−0.12	0.06
24	147.67	−0.33	−0.12	0.05	141.63	−0.35	−0.17	0.03
26	150.22	−0.37	−0.13	0.07	143.28	−0.38	−0.17	0.05
28	152.94	−0.42	−0.13	0.09	145.05	−0.41	−0.16	0.08
30	155.76	−0.53	−0.20	0.05	146.73	−0.65	−0.36	−0.09
32	158.75	−0.63	−0.25	0.03	148.73	−0.68	−0.35	−0.06
34	161.91	−0.70	−0.27	0.03	150.82	−0.73	−0.36	−0.04
36	165.16	−0.82	−0.34	−0.01	152.97	−0.84	−0.41	−0.06
38	168.61	−0.88	−0.34	0.02	155.27	−0.89	−0.42	−0.04
40	172.16	−0.96	−0.36	0.03	157.63	−1.00	−0.46	−0.05
42	175.78	−1.10	−0.43	0.00	160.09	−1.11	−0.51	−0.06
44	179.50	−1.24	−0.50	−0.05	162.60	−1.24	−0.59	−0.11
46	183.58	−1.14	−0.32	0.17	165.29	−1.30	−0.58	−0.07
48	187.30	−1.49	−0.59	−0.07	168.08	−1.35	−0.56	−0.02
50	191.37	−1.59	−0.60	−0.05				

Table 9 (continued)

J	$^{Q}S_{1,1,0}(J)$ obs.	obs. − calc. I	II	III	$^{Q}S_{1,3,0}(J)$ obs.	obs. − calc. I	II	III	$^{Q}Q_{1,1,2}(J)$ obs.	obs. − calc. I	II	III
2	6.26	0.02	0.03	0.03								
4	9.82	0.02	0.03	0.03	10.27	0.01	0.02	0.03	144.35	0.00	0.01	−0.07
6	13.28	−0.08	−0.07	−0.07	13.93	−0.07	−0.05	−0.04	144.81	0.00	0.02	−0.06
8	16.85	−0.06	−0.05	−0.05	17.73	−0.01	0.02	0.03	145.45	−0.01	0.03	−0.05
10	20.49	0.03	0.04	0.04	21.46	−0.02	0.02	0.02	146.31	0.03	0.08	0.01
12	24.02	0.01	0.03	0.03	25.16	−0.06	−0.01	0.00	147.30	0.00	0.08	0.01
14	27.47	−0.08	−0.06	−0.05	29.00	0.05	0.10	0.11	148.42	−0.09	0.02	−0.04
16	31.05	−0.03	0.00	0.00	32.57	−0.10	−0.04	−0.03	149.84	−0.08	0.06	0.01
18	34.57	−0.03	0.00	0.00	36.37	−0.02	0.05	0.07	151.38	−0.13	0.05	0.00
20	38.06	−0.06	−0.03	−0.03	40.12	0.03	0.11	0.13	153.12	−0.17	0.04	0.01
22	41.59	−0.04	0.00	0.00	43.68	−0.10	−0.01	0.01	155.08	−0.18	0.08	0.06
24	45.03	−0.11	−0.05	−0.05	47.33	−0.12	−0.02	0.00	157.15	−0.25	0.05	0.05
26	48.56	−0.07	−0.01	−0.01					159.44	−0.28	0.08	0.09
28	52.04	−0.08	−0.01	−0.01					161.87	−0.33	0.08	0.11
30	55.51	−0.09	0.00	−0.01					164.52	−0.32	0.15	0.19
32	58.98	−0.09	0.01	0.00								
34	62.46	−0.07	0.04	0.03								
36	65.89	−0.10	0.03	0.02								
38	69.18	−0.26	−0.11	−0.12								
40	72.71	−0.17	0.00	−0.01								
42	76.03	−0.28	−0.10	−0.11								

Table 9 (continued)

J	$^{Q}S_{1,1,0}(J)$ obs.	obs. − calc. I	II	III	$^{Q}S_{1,3,0}(J)$ obs.	obs. − calc. I	II	III	$^{Q}Q_{1,1,2}(J)$ obs.	obs. − calc. I	II	III
3									144.11	−0.01	−0.01	−0.09
5					11.82	−0.05	−0.03	−0.03	144.37	−0.05	−0.03	−0.11
7	14.93	−0.01	−0.01	0.00	15.48	−0.06	−0.03	−0.03	144.81	−0.05	−0.02	−0.10
9	18.40	−0.06	−0.04	−0.04	19.26	0.03	0.06	0.07	145.41	−0.04	0.00	−0.07
11	21.99	0.03	0.04	0.04	22.89	−0.07	−0.03	−0.01	146.23	0.00	0.06	0.01
13	25.43	−0.04	−0.02	−0.02	26.66	−0.05	−0.01	0.02	147.17	−0.06	0.03	−0.01
15	28.90	−0.07	−0.05	−0.04	30.35	−0.12	−0.06	−0.03	148.40	−0.07	0.04	0.04
17	32.46	−0.01	0.02	0.02	34.13	−0.07	−0.01	0.02	149.83	−0.14	0.01	0.03
19	35.97	0.01	0.05	0.05	37.83	−0.08	−0.01	0.02	151.53	−0.18	0.00	0.06
21	39.45	0.01	0.05	0.05	41.66	−0.07	0.14	0.17	153.40	−0.26	−0.04	0.04
23	42.92	0.00	0.04	0.04					155.61	−0.20	0.06	0.16
25	46.35	−0.05	0.01	0.01					158.01	−0.14	0.16	0.29
27	49.83	−0.04	0.03	0.03								
29	53.34	0.01	0.08	0.08								
31	56.66	−0.13	−0.04	−0.05								
33	60.20	−0.04	0.06	0.05								
35	63.62	−0.07	0.04	0.04								
37	67.06	−0.07	0.06	0.05								
39	70.51	−0.06	0.09	0.08								
41												
43	77.33	−0.10	0.09	0.08								

transitions. Therefore they appear as transitions with Raman selection rules, with series corresponding to pure rotational Raman lines,

$$^{Q}S_{l,v,0}(J): |v, J+2, l, \gamma\rangle \leftarrow |v, J, l, \gamma\rangle$$

and vibration-rotation series,

$$^{Q}Q_{l,v,2}(J): |v+2, J, l, \gamma\rangle \leftarrow |v, J, l, \gamma\rangle$$
$$^{S}Q_{l,v,2}(J): |v+2, J, l+2, \gamma\rangle \leftarrow |v, J, l, \gamma\rangle$$

γ has been omitted from the series designation, since spin statistics establishes a unique relation between the parities of J and γ. The three sets of deviations between observed and calculated frequencies correspond to calculations based on the three parameter sets of Table 7.

Thus calculation I was based on the same parameters as used in obtaining Table 8. It is noted that the pure rotational energy differences, the $^{Q}S_{l,v,0}$-series, are reproduced excellently for low J-values, whereas increasingly negative deviations appear with increasing J (the large deviations of the first members of the $^{Q}S_{l,2,0}$-series seem to be caused by extreme experimental errors). The discrepancies at high J-values are even more pronounced when the vibrational differences, the $Q_{v,l,2}$-series, are considered. However, it may also be noticed that the $^{Q}Q_{0,0,2}$- and the $^{S}Q_{0,0,2}$-deviations are nearly identical at given J. This means that the splittings of the $v = 2$ states, enhanced by l-type resonance, are excellently reproduced. Similarly we may compare the deviations in the two $^{Q}Q_{1,1,2}$-series exhibiting the effects of l-type doubling (the two $^{S}Q_{1,1,2}$-series, further split by l-type resonance, are not experimentally available). Thus we observe that these effects are reproduced as well.

The deviations in the rotational series are of exactly the order of magnitude expected for centrifugal distortion displacements (Sect. 5.5). Since the precision of measurement is sufficient to make these perturbations clearly visible, it was decided to investigate the centrifugal distortion operator discussed in Sect. 4.8.2.3 more closely.

As suggested earlier we apply ρ-independent centrifugal distortion constants which are derived from the partial derivatives, $B_{gg}^{(1)}$, at $\rho = 0$. $B_{zz}^{(1)}$ should be given special consideration, however. From Table 3 we find that the derivatives can be conveniently expressed using $D_J = 4\,B^3/\omega_1^2$,

$$\begin{aligned} B_{xx}^{(1)} &= B_{yy}^{(1)} = -(2\,\omega_1 D_J)^{1/2} \\ B_{\rho\rho}^{(1)} &= \rho^2\,B_{zz}^{(1)} = -36\,(2\,\omega_1 D_J)^{1/2} \end{aligned} \qquad (5.41)$$

Notice, that $B_{zz}^{(1)}$ has not been approximated by a constant. We proceed most easily by applying the classical method. Thus the equation corresponding to equation (4.67) reads

$$\frac{1}{hc}\frac{\partial H}{\partial q_1} = \omega_1 q_1 - (2\,\omega_1 D_J)^{1/2}\left[J_x^2 + J_y^2 + 36\left(J_\rho^2 + \left(J_z^2 - \frac{1}{4}\right)\Big/\rho^2\right)\right]$$

from which we find the perturbation

$$H_{\text{C.D.}} = -D_J \left[J^2 - J_z^2 + 36 \left(J_\rho^2 + \left(J_z^2 - \frac{1}{4} \right) / \rho^2 \right) \right]^2 \tag{5.42}$$

This result indicates that the large deviations of the $Q_{0,0,2}$-series can be explained by our previous neglect of the product term

$$-72\, D_J \left(J_\rho^2 + \left(J_z^2 - \frac{1}{4} \right) / \rho^2 \right) (J^2 - J_z^2)$$

which reduces the effective rotational constants by

$$\Delta \mathrm{B_{eff}} = -72\, D_J \, \langle v, l | J_\rho^2 + \left(J_z^2 + \frac{1}{4} \right) / \rho^2 \, | v, l \rangle \tag{5.43}$$

For $v = 0, 2, 4$ and 6 this amounts to about $-9 \cdot 10^{-5}$, $-3.3 \cdot 10^{-4}$, $-6.4 \cdot 10^{-4}$ and $-1.02 \cdot 10^{-3}\ \mathrm{cm}^{-1}$, and for a $Q_{0,0,2}(40)$-difference we may expect a centrifugal distortion displacement of $-0.40\ \mathrm{cm}^{-1}$. Thus centrifugal distortion accounts for only a little less then half the deviations, -0.96 and $-0.100\ \mathrm{cm}^{-1}$ at $J = 40$.

In calculation II the centrifugal distortion effects from the operator of Eq. (5.42) were rigorously evaluated. It is seen that the rotational series, ${}^{Q}S_{l,\,v,\,0}$, are very well reproduced. This is also the case for the ${}^{Q}Q_{1,1,2}$-series (at the rather low J-values included) whereas the $Q_{0,0,2}$-series still show significant systematic deviations.

In the ordering scheme discussed in Sect. 4.8 centrifugal distortion and relaxation effects are comparable in magnitude. Having found significant centrifugal distortion displacements, it seems probable that the remaining deviations correspond to relaxation effects.

To investigate this possibility, we introduce a dimensionless relaxation parameter, k, such that $V_2^{(1)}$ of Eq. (5.16) is given by

$$V_2^{(1)} = k\, (\omega_1^3 / 8\, B)^{1/2} \tag{5.44}$$

The corresponding corrections, $\Delta B_{gg}^{\mathrm{rel}}$, calculated from Eq. (4.69) and Table 3 are,

$$\begin{aligned} \Delta B_{xx}^{\mathrm{rel.}} &= Bk\rho^2\, 2 f_2^{-1} f_1^{-1/2}, & \Delta B_{yy}^{\mathrm{rel.}} &= Bk\rho^2\, 9 f_3^{-2} f_1^{-1/2} \\ \Delta B_{zz}^{\mathrm{rel.}} &= Bk\rho^2\, 18 f_0^{-1} f_1^{-1/2}, & \Delta B_{\rho\rho}^{\mathrm{rel.}} &= Bk\rho^2\, 36 f_1^{-5/2} \end{aligned} \tag{5.45}$$

which were used in determining the expansion coefficients of the ρ-dependent terms, Table 6.

A rough estimate of the effect of k in relation to the deviations of the $Q_{0,0,2}$-series is found by considering the change in effective rotational constants. The difference, δB_{eff}, between the $v = 0$ and the $v = 2$ constants, is changed by

$$\Delta \delta B_{\mathrm{eff}} = Bk\, \delta \langle \rho^2 \rangle = Bk\, 0.24, \tag{5.46}$$

so aiming at a shift in calculated frequency of about -0.40 cm^{-1} at $J = 40$ we find that $k = -0.0024$ might be a feasible value.

In calculation III a satisfactory fit to all the energy differences could be obtained using the value $k = -0.002$. This corresponds to a relaxation of the bond length of

$$\Delta R = 0.0012\, \rho^2\, R \tag{5.47}$$

or a lengthening of about 0.1% at 50°. This relaxation is so small that, from a chemical point of view, we would say that the C,C bond length of C_3 is independent of the bending angle. The significance of the relaxation may also be questioned since a change in the anharmonicity parameter, a, [Eq. (5.12)] from 2 Å^{-1} to 3.5 Å^{-1} gives an equally satisfying fit as obtained by introducing k. Thus a and k are highly correlated and consequently it is impossible to make conclusive statements as to the importance of relaxation. Although $a = 3.5$ Å^{-1} is somewhat outside the usual range of values, we must recall that the bonding in C_3 is unusual as well.

5.8 Conclusion

The main object of presenting this example was to illustrate how the theory is applied in practice and to show that the theoretical predictions are in agreement with experiment values. Hopefully the rather detailed evaluation has served the first

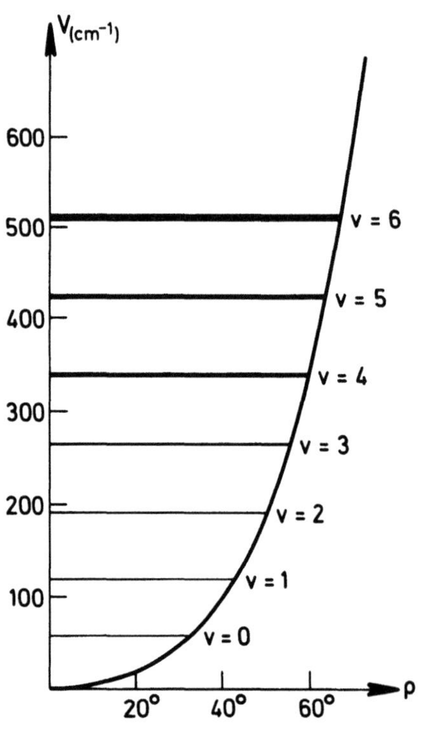

Fig. 2. Potential function of the bending in C_3 with energy levels up to $v = 6$. The thickness of level lines indicate the magnitude of l-splittings

purpose. The latter goal has been reached if we may agree that the deviations of calculation III in Table 9 are accidental. This should be possible since the weak indications of systematic errors in some of the series may just as well be ascribed to errors of measurement.

From the calculations we have in addition gained some new information about C_3. We may conclude that the equilibrium bond length of C_3 is

$$r_e = 1.2945 \pm 0.0005 + (0.0016 \pm 0.0010)\,\rho^2 \text{ Å}, \tag{5.48}$$

where the uncertainties were estimated from the scattering of B-values in Table 7 with the preceding discussion of relaxation in mind.

Also the effective potential function is very well determined. The three sets of V_n constants in Table 7 varies considerably, but this only demonstrates the correlation between the parameters. When the corresponding potential curves are examined, they are found to be so close that Fig. 2 can depict them all within the thickness of drawing.

Calculations have not been made for the bending states with $\nu = 4$ and 6, although there are experimental data available. The treatment could easily be extended to these levels as well, but the additional information that might result is scarcely essential. It would be much more interesting if more experimental data were available for excited states of the stretching modes. At present only one number can be extracted from the available data[83)] for comparison with the theory. That is $\alpha_1^{(B)} = 0.0048$ cm^{-1} obtained from the effective rotational constants of the ν_1 and $2\,\nu_1$ states. This agrees very well with the value 0.00496 cm^{-1} estimated from expectation values of the g.s.

6 References

1. Wilson, E. B., Howard, J. B.: J. Chem. Phys. *4*, 260 (1936)
2. Wilson, E. B., Decius, J. C., Cross, P. C.: Molecular vibrations. New York: McGraw-Hill 1955
3. Wilson, E. B.: J. Chem. Phys. *7*, 1047 (1939) and *9*, 76 (1941)
4. Amat, G., Nielsen, H. H., Tarrago, G.: Rotation vibration of polyatomic molecules. New York: Marcel Dekker, Inc. 1971
5. Mills, I. M.: Vibration-rotation structure in asymmetric- and symmetric-top molecules. In: Molecular spectroscopy: Modern research. Rao, K. N., Mathews, C. W. (ed.). New York-London: Academic Press 1972
6. Hoy, A. R., Mills, I. M., Strey, G.: Mol. Phys. *24*, 1265 (1972)
7. Quade, C. R.: J. Chem. Phys. *64*, 2783 (1976)
8. Dreizler, H.: Fortschr. Chem. Forsch. *10/1*, 59 (1968)
9. Dreizler, H.: Rotational spectra of molecules with two internal degrees of freedom. In: See Ref. [5])
10. Thorson, W., Nakagawa, I.: J. Chem. Phys. *33*, 994 (1960)
11. Dixon, R. N.: Trans. Far. Soc. *60*, 1363 (1964)
12. Johns, J. W. C.: Can. J. Phys. *45*, 2639 (1967)
13. Carreira, L. A., Carter, R. O., Durig, J. R., Lord, R. C., Milionis, C. C.: J. Chem. Phys. *59*, 1028 (1973)
14. Durig, J. R., Kalasinsky. K. S., Kalasinsky, V. F.: J. Chem. Phys. *69*, 918 (1978)
15. Duckett, J. A., Robiette, A. G., Mills, I. M.: J. Mol. Spectrosc. *62*, 19, 34 (1976) and *63*, 249 (1976)
16. Shinkle, N. L., Coon, J. B.: J. Mol. Spectrosc. *40*, 217 (1971)
17. Hougen, J. T., Bunker, P. R., Johns, J. W. C.: J. Mol. Spectrosc. *34*, 136 (1970)
18. Bunker, P. R., Stone, J. M. R.: J. Mol. Spectrosc. *41*, 310 (1972)
19. Bunker, P. R., Landsberg, B. M.: J. Mol. Spectroscopy *67*, 374 (1977)
20. Stone, J. M. R.: J. Mol. Spectrosc. *54*, 1 (1975)
21. Bunker, P. R., Landsberg, B. M., Winnewisser, B. P.: J. Mol. Spectrosc. *74*, 9 (1979)
22. Weber, W. H., Ford, G. W.: J. Mol. Spectrosc. *63*, 445 (1976)
23. Moule, D. C., Rao, Ch. V. S. R.: J. Mol. Spectrosc. *45*, 120 (1973)
24. Papoušek, D., Stone, J. M. R., Špirko, V.: J. Mol. Spectrosc. *48*, 17 (1973)
25. Danielis, V., Papoušek, D., Špirko, V., Horak, M.: J. Mol. Spectrosc. *54*, 339 (1975)
26. Kreglewski, M.: J. Mol. Spectrosc. *72*, 1 (1978)
27. Henderson, G., Ewing, G. E.: J. Chem. Phys. *59*, 2280 (1973)
28. Istomin, V. A., Stepanov, N. F., Zhilinskii, B. I.: J. Mol. Spectrosc. *67*, 265 (1977)
29. Malloy, T. B.: J. Mol. Spectrosc. *44*, 504 (1972)
30. Malloy, T. B., Lafferty, W. J.: J. Mol. Spectrosc. *54*, 20 (1975)
31. Villarreal, J. R., Bauman, L. E., Laane, J., Harris, W. C., Bush, S. F.: J. Chem. Phys. *63*, 3727 (1975)
32. Bauder, A., Mathier, E. Meyer, R., Ribeaud, M., Günthard, Hs. H.: Mol. Phys. *15*, 597 (1968)
33. Carreira, L. A.: J. Chem. Phys. *62*, 3851 (1975)
34. Susskind, J.: J. Chem. Phys. *53*, 2492 (1970)
35. Bauder, A., Günthard, Hs. H.: J. Mol. Spectrosc. *60*, 290 (1976)
36. Kirtman, B.: J. Chem. Phys. *37*, 2516 (1962)
37. Hoy, A. R., Bunker, P. R.: J. Mol. Spectrosc. *52*, 439 (1974), *54*, 165 (1975) and *59*, 159 (1976)
38. Špirko, V., Stone, J. M. R., Papoušek, D.: J. Mol. Spectrosc. *60*, 159 (1976)
39. Eckart, C.: Phys. Rev. *47*, 552 (1935)
40. Sayvetz, A.: J. Chem. Phys. *7*, 383 (1939)
41. Iijima, T., Tsuchiya, S.: J. Mol. Spectrosc. *44*, 88 (1972)
42. Fleming, J. W., Banwell, C. N.: J. Mol. Spectrosc. *31*, 318 and 378 (1969)
43. Quade, C. R.: J. Chem. Phys. *44*, 2512 (1966)
44. Freed, K. F., Lombardi, J. R.: J. Chem. Phys. *45*, 591 (1966)

45. Darling, B. T., Nadeau, G.: J. Mol. Spectrosc. *11*, 200 (1963)
46. Celles, M. de, Darling, B. T.: J. Mol. Spectrosc. *29*, 66 (1969)
47. Sarka, K.: J. Mol. Spectrosc. *38*, 545 (1971)
48. Brand, J. C. D., Rao, Ch. V. S. R.: J. Mol. Spectrosc. *61*, 360 (1976)
49. Allen, H. C., Cross, P. C.: Molecular vib-rotors. New York: Wiley 1963
50. Margenau, H., Murphy, G. M.: The mathematics of chemistry and physics. New York–London–Toronto: D. Van Nostrand 1956
51. Kroto, H. W.: Molecular rotation spectra. London–New York–Sydney–Toronto: John Wiley & Sons 1975
52. Schutte, C. J. H.: The theory of molecular spectroscopy, Vol. I. Amsterdam–Oxford–New York: North Holland-American Elsevier Publ. Co. 1976
53. Kuchitsu, K., Morino, Y.: Bull. Chem. Soc. Japan *38*, 805 and 814 (1965)
54. Meyer, R., Günthard, Hs. H.: J. Chem. Phys. *49*, 1510 (1968)
55. Polo, S. R.: J. Chem. Phys. *24*, 1133 (1956)
56. Malhiot, R. J., Ferigle, S. M.: J. Chem. Phys. *22*, 717 (1954)
57. Amat, G., Henry, L.: Cah. Phys. *12*, 273 (1958)
58. Watson, J. K. G.: Mol. Phys. *15*, 479 (1968)
59. Watson, J. K. G.: Mol. Phys. *19*, 465 (1970)
60. Podolsky, B.: Phys. Rev. *32*, 812 (1928)
61. Cross, P. C., Hainer, R. M., King, G. W.: J. Chem. Phys. *12*, 210 (1944)
62. Pariseau, M. A., Suzuki, I., Overend, J.: J. Chem. Phys. *42*, 2335 (1965)
63. Meal, J. H., Polo, S. R.: J. Chem. Phys. *24*, 1119 (1956)
64. Nielsen, H. H.: Rev. Mod. Phys. *23*, 90 (1951)
65. Dowling, J. M.: J. Mol. Spectrosc. *6*, 550 (1961)
66. Wollrab, J. E.: Rotational spectra and molecular structure. New York–London: Academic Press 1967
67. Gordy, W., Cook, R. L.: Microwave molecular spectra. New York–London–Sydney–Toronto: Interscience Publishers, John Wiley & Sons 1970
68. Newton, R. R., Thomas, L. H.: J. Chem. Phys. *16*, 310 (1948)
69. Meyer, R., Wilson, E. B.: J. Chem. Phys. *53*, 3969 (1970)
70. Ribeaud, M., Bauder, A., Günthard, Hs. H.: Mol. Phys. *23*, 235 (1972)
71. Nösberger, P., Bauder, A., Günthard, Hs. H.: Chem. Phys. *4*, 196 (1974)
72. Jørgensen, F., Pedersen, Th.: Mol. Phys. *27*, 33 (1974)
73. Jørgensen, F. Pedersen, Th.: Mol. Phys. *27*, 959 (1974)
74. Jørgensen, F. Pedersen, Th., Chedin, A.: Mol. Phys. *30*, 1377 (1975)
75. Lin, C. C., Swalen, J. D.: Rev. Mod. Phys. *31*, 841 (1959)
76. Sørensen, G. O.: Unpubl. results
77. Pickett, H. M.: J. Chem. Phys. *56*, 1715 (1972)
78. Lees, R. M.: J. Chem. Phys. *57*, 824 (1972)
79. Yamada, K., Winnewisser, M.: Z. Naturforsch. A *31*, 139 (1976)
80. Hansen, C. F., Pearson, W. E.: Can. J. Phys. *51*, 751 (1973)
81. Hansen, C. F., Henderson, B. J., Pearson, W. E.: J. Chem. Phys. *60*, 754 (1974)
82. Gausset, L., Herzberg, G., Lagerqvist, A., Rosen, B.: Astrophys. J. *142*, 45 (1967)
83. Merer, J.: Can. J. Phys. *45*, 4103 (1967)
84. Weltner Jr., W., McLeod Jr., D.: J. Chem. Phys. *40*, 1305 (1964)
85. Louck, J. D., Schaffer, W. H.: J. Mol. Spectrosc. *4*, 285 (1960)

Received November 28, 1978

Author Index Volumes 26–82

The volume numbers are printed in italics

Large Amplitude Motion in Molecules I

1979. 30 figures. Approx. 190 pages.
(Topics in Current Chemistry, Volume 81)
ISBN 3-540-09310-9

Contents/Information:

H. Frei, A. Bauder, H.-H. Günthard
The Isometric Group of Nonrigid Molecules

The method presented here has been motivated by the desire to find a method which starts from the geometrical description of nuclear configurations and replaces the feasibility concept by rigorous mathematical definitions.
(61 references)

O. Bastiansen, K. Kveseth, H. Møllendal
Structure of Molecules with Large Amplitude Motion as Determined from Electron-Diffraction Studies in the Gas Phase

The large amplitude motion that sometimes takes place within the molecule should above all be studied in the gas phase.
(413 references)

Springer-Verlag
Berlin
Heidelberg
New York

Reactivity and Structure

Concepts in Organic Chemistry

Editors: K. Hafner, J.-M. Lehn, C. W. Rees, P. v. Ragué Schleyer, B. M. Trost, R. Zahradník

This series will not only deal with problems of the reactivity and structure of organic compounds but also consider synthetical-preparative aspects.
Suggestions as to topics will always be welcome.

Volume 1: J. Tsuji
Organic Synthesis
by Means of Transition Metal Complexes
A Systematic Approach
1975. 4 tables. IX, 199 pages
ISBN 3-540-07227-6

Volume 2: K. Fukui
Theory of Orientation and Stereoselection
1975. 72 figures, 2 tables. VII, 134 pages
ISBN 3-540-07426-0

Volume 3: H. Kwart, K. King
d-Orbitals in the Chemistry of Silicon, Phosphorus and Sulfur
1977. 4 figures, 10 tables. VIII, 220 pages
ISBN 3-540-07953-X

Volume 4: W. P. Weber, G. W. Gokel
Phase Transfer Catalysis in Organic Synthesis
1977. 100 tables. XV, 280 pages
ISBN 3-540-08377-4

Volume 5: N. D. Epiotis
Theory of Organic Reactions
1978. 69 figures, 47 tables. XIV, 290 pages
ISBN 3-540-08551-3

Volume 6: M. L. Bender, M. Komiyama
Cyclodextrin Chemistry
1978. 14 figures, 37 tables. X, 96 pages
ISBN 3-540-08577-7

Volume 7: D. I. Davies, M. J. Parrott
Free Radicals in Organic Synthesis
1978. 1 figure. XII, 169 pages
ISBN 3-540-08723-0

Volume 8: C. Birr
Aspects of the Merrifield Peptide Synthesis
1978. 62 figures, 6 tables. VIII, 102 pages
ISBN 3-540-08872-5

Volume 9: J. R. Blackborow, D. Young
Metal Vapour Synthesis in Organometallic Chemistry
1979. 36 figures, 30 tables. Approx. 220 pages
ISBN 3-540-09330-3

Springer-Verlag
Berlin
Heidelberg
New York

Batch number: 10620881

Printed by Printforce, the Netherlands